Math
FOR THE TRADES

Math

FOR THE TRADES

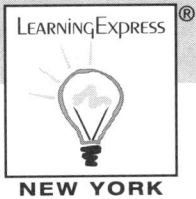

LEARNINGEXPRESS®

NEW YORK

Library of Congress Cataloging-in-Publication Data:
  Math for the trades / LearningExpress.—1st ed.
      p. cm.
  ISBN 1-57685-515-5
    1. Mathematics. I. LearningExpress (Organization)
  QA39.3.M378 2004
  513'.14—dc22

                                        2004001656

Printed in the United States of America

9 8 7 6 5 4 3 2 1

First Edition

ISBN 1-57685-515-5

For more information or to place an order, contact LearningExpress at:
    55 Broadway
    8th Floor
    New York, NY 10006

Or visit us at:
    www.learnatest.com

Contributors

Kristin Davidson is a math teacher at The Bishop's School in San Diego, California.

Ashley Clark is a former math and science teacher from San Diego, California. She is currently pursuing her M.D. from The University of Vermont.

Melinda Grove taught middle school math for seven years in Connecticut and has been an adjunct math professor for three years. She is currently a math consultant for several publications.

Lara Bohlke has a Bachelor's Degree in Mathematics and a Master's Degree in Mathematics Education. She has been a math teacher since 1989 and has taught eighth grade through college level mathematics.

Colleen Schultz is a math teacher and teacher mentor in Vestal, New York. She is a contributing math writer for *501 Math Problems* and *Just in Time Algebra*.

Catherine V. Jeremko is a math teacher and expert math reviewer from Vestal, New York. She is the author of *Just in Time Math*.

Contents

Math

FOR THE TRADES

1 How to Use This Book

A career in the trades can be very rewarding. Whether you have just started, or have worked for several years, having strong math skills is important for success on the job. You may even have to take a math competency test to be hired for some jobs. Maybe you haven't used your math skills in a while, or maybe you need to improve your math skills to move on to a better job, or simply to succeed at the job you are doing. Whatever the situation, by making the commitment to practice your math skills, you are promising yourself increased success and marketability. With over 200 on-the-job practice questions in arithmetic, measurement, basic algebra, basic geometry, word problems, and data analysis, this book is designed just for you!

▶ ABOUT *MATH FOR THE TRADES*

You should carefully read this chapter so you can grasp effective strategies and learn to make the most of the lessons in this book. When you finish this chapter, take the 50-question pretest. Don't worry if you haven't studied math in a while. Your score on the pretest will help you gauge your current level of math skills and show you which lessons you need to review the most. After you take the pretest, you can refer to the answer explanations to see exactly how to solve each of the questions. The pretest begins with basic-level questions, and they gradually increase in difficulty. All of the questions on the pretest and throughout this book are word problems set in the context of work-related problems. The questions are meant to reflect the types of math problems that occur in the trade workplace. Some of these jobs include:

▶ retail (cashier, stockperson, salesperson)
▶ construction (carpenter, electrician)
▶ landscaping
▶ food service (cook, buyer, server)
▶ customer service (telemarketing, front desk, delivery person)
▶ home repair (painters, carpenters, carpet layers, movers, housecleaners, plumbers)

Before you take the pretest, let's review some basic math strategies.

▶ MATH STRATEGIES

These suggestions are tried and true. You may use one or all of them. Or, you may decide to pick and choose the combination that works best for you.

▶ It is best not to work in your head! Use scratch paper to take notes, draw pictures, and calculate. Although you might think that you can solve math questions more quickly in your head, that's a good way to make mistakes. Instead, write out each step.

▶ Before you begin to make your calculations, read a math question in chunks rather than straight through from beginning to end. As you read each chunk, stop to think about what it means. Then make notes or draw a picture to represent that chunk.

▶ When you get to the actual question, circle it. This will keep you more focused as you solve the problem.

▶ Glance at the answer choices for clues. If they are fractions, you should do your work in fractions; if they are decimals, you should work in decimals, etc.

▶ Develop a plan of attack to help you solve the problem. When you get your answer, reread the circled question to make sure you have answered it. This helps avoid the careless mistake of answering the wrong question.

▶ Always check your work after you get an answer. You may have a false sense of security when you get an answer that matches one of the multiple-choice answers. It could be right, but you should always check your work. Remember to:
  ■ Ask yourself if your answer is reasonable, if it makes sense.
  ■ Plug your answer back into the problem to make sure the problem holds together.
  ■ Do the question a second time, but use a different method.

■ Approximate when appropriate. For example:

$5.98 + $8.97 is a little less than $15. (Add: $6 + $9)

.9876 × 5.0342 is close to 5. (Multiply: 1 × 5)

▶ Skip questions that you find difficult and come back to them later. Make a note about them so you can find them quickly.

▶ REVIEW AND PRACTICE

Once you have completed the pretest in **Chapter 2** and reviewed all the answer explanations, you are ready to move on.

Chapter 3 covers the basic elements of arithmetic. You will learn about numbers, symbols, operations, fractions, decimals, percents, averages, and square roots.

Chapter 4 is a review of measurement skills. You will learn about using different measurement systems, performing mathematical operations with units of measurement, and converting between different units.

Chapter 5 covers basic algebra skills. You will become familiar with variables, cross multiplying, algebraic fractions, reciprocal rules, and exponents.

Chapter 6 reviews the basics of geometry. You will study the properties of angles, lines, polygons, triangles, and circles, as well as the formulas for area, volume, and perimeter.

Chapter 7 is a thorough review of word problems and data analysis questions. It may sound difficult, but it is not. This lesson will show you how to set up and solve word problems, and understand graphs, charts, tables, and diagrams with confidence.

Chapters 3–7 each have sample problems within the lesson, but when you finish reading each lesson, you will have a chance to solve 15 practice questions on the topics you just reviewed. The questions increase in difficulty, but each question includes a thorough answer explanation to reinforce what you just learned.

When you have completed each lesson and practice set, you are ready to see how much you have improved. **Chapter 8** includes a 100-question post-test covering the same types of math you will have studied in the previous chapters. Again, the first questions are more basic, and they get more difficult. If you don't understand a question, remember the post-test is followed by answer explanations to help you. When you are done, compare your score on the pretest to your score on the post-test and see how much you have improved.

Good luck!

WORKING BACKWARDS

You can frequently solve a word problem by plugging the answer choices back into the text of the problem to see which one fits all the facts stated in the problem. The process is faster than you think because you will probably only have to substitute one or two answers to find the right one. This approach works only when:

- all of the answer choices are numbers.
- you are asked to find a simple number, not a sum, product, difference, or ratio.

Here's what to do:

1. Look at all the answer choices and begin with the one in the middle of the range. For example, if the answers are 14, 8, 2, 20, and 25, begin by plugging 14 into the problem.
2. If your choice doesn't work, eliminate it. Determine if you need a bigger or smaller answer.
3. Plug in one of the remaining choices.
4. If none of the answers work, you may have made a careless error. Begin again or look for your mistake.

Example:

Juan sold $\frac{1}{3}$ of the books in the store during the morning shift. On the evening shift, Marcella sold $\frac{3}{4}$ of the remaining stock, which left 10 books. How many books were there to begin with?

a. 60
b. 80
c. 90
d. 120

Starting with the middle answer, let's assume there were 90 books to begin with:

Since Juan sold $\frac{1}{3}$ of them, that means he sold 30 ($\frac{1}{3} \times 90 = 30$), leaving 60 of them ($90 - 30 = 60$). Marcella then sold $\frac{3}{4}$ of the 60 books, or 45 of them ($\frac{3}{4} \times 60 = 45$). That leaves 15 books ($60 - 45 = 15$).

The problem states that there were 10 books left, and using this answer, we ended up with 15 of them. That indicates that we started with too big a number. Thus, 90 and 120 are both wrong! With only two choices left, let's use common sense to decide which one to try. The next lower answer is only a little smaller than 90 and may not be small enough. So, let's try 60:

Since Juan sold $\frac{1}{3}$ of them, that means he sold 20 ($\frac{1}{3} \times 60 = 20$), leaving 40 of them ($60 - 20 = 40$). Marcella then sold $\frac{3}{4}$ of the 40 books, or 30 of them ($\frac{3}{4} \times 40 = 30$). That leaves 10 books ($40 - 30 = 10$).

Because this result of 10 books remaining agrees with the problem, the correct answer is **a**.

Pretest

This pretest is designed to gauge your math skills before you review the lessons in this book. Perhaps you have encountered some of these types of math problems before, whether in a classroom or at work. If so, you will probably feel at ease answering some of the following questions. However, there may be other questions that you find difficult. This test will help to pinpoint any weaknesses you may have so you can study the lessons that cover the skills you need to work on.

There are 50 multiple-choice questions in the pretest. Take as much time as you need to answer each one. If this is your book, you may simply fill in the correct answer on the answer sheet on page 6. If the book does not belong to you, use a separate sheet of paper to record your answers, numbering 1 through 50. You may use a calculator, but your practice will be more effective if you try to solve the problems on your own. When you finish the test, use the answer explanations to check your results.

▶ **ANSWER SHEET**

1.	ⓐ	ⓑ	ⓒ	ⓓ	26.	ⓐ	ⓑ	ⓒ	ⓓ
2.	ⓐ	ⓑ	ⓒ	ⓓ	27.	ⓐ	ⓑ	ⓒ	ⓓ
3.	ⓐ	ⓑ	ⓒ	ⓓ	28.	ⓐ	ⓑ	ⓒ	ⓓ
4.	ⓐ	ⓑ	ⓒ	ⓓ	29.	ⓐ	ⓑ	ⓒ	ⓓ
5.	ⓐ	ⓑ	ⓒ	ⓓ	30.	ⓐ	ⓑ	ⓒ	ⓓ
6.	ⓐ	ⓑ	ⓒ	ⓓ	31.	ⓐ	ⓑ	ⓒ	ⓓ
7.	ⓐ	ⓑ	ⓒ	ⓓ	32.	ⓐ	ⓑ	ⓒ	ⓓ
8.	ⓐ	ⓑ	ⓒ	ⓓ	33.	ⓐ	ⓑ	ⓒ	ⓓ
9.	ⓐ	ⓑ	ⓒ	ⓓ	34.	ⓐ	ⓑ	ⓒ	ⓓ
10.	ⓐ	ⓑ	ⓒ	ⓓ	35.	ⓐ	ⓑ	ⓒ	ⓓ
11.	ⓐ	ⓑ	ⓒ	ⓓ	36.	ⓐ	ⓑ	ⓒ	ⓓ
12.	ⓐ	ⓑ	ⓒ	ⓓ	37.	ⓐ	ⓑ	ⓒ	ⓓ
13.	ⓐ	ⓑ	ⓒ	ⓓ	38.	ⓐ	ⓑ	ⓒ	ⓓ
14.	ⓐ	ⓑ	ⓒ	ⓓ	39.	ⓐ	ⓑ	ⓒ	ⓓ
15.	ⓐ	ⓑ	ⓒ	ⓓ	40.	ⓐ	ⓑ	ⓒ	ⓓ
16.	ⓐ	ⓑ	ⓒ	ⓓ	41.	ⓐ	ⓑ	ⓒ	ⓓ
17.	ⓐ	ⓑ	ⓒ	ⓓ	42.	ⓐ	ⓑ	ⓒ	ⓓ
18.	ⓐ	ⓑ	ⓒ	ⓓ	43.	ⓐ	ⓑ	ⓒ	ⓓ
19.	ⓐ	ⓑ	ⓒ	ⓓ	44.	ⓐ	ⓑ	ⓒ	ⓓ
20.	ⓐ	ⓑ	ⓒ	ⓓ	45.	ⓐ	ⓑ	ⓒ	ⓓ
21.	ⓐ	ⓑ	ⓒ	ⓓ	46.	ⓐ	ⓑ	ⓒ	ⓓ
22.	ⓐ	ⓑ	ⓒ	ⓓ	47.	ⓐ	ⓑ	ⓒ	ⓓ
23.	ⓐ	ⓑ	ⓒ	ⓓ	48.	ⓐ	ⓑ	ⓒ	ⓓ
24.	ⓐ	ⓑ	ⓒ	ⓓ	49.	ⓐ	ⓑ	ⓒ	ⓓ
25.	ⓐ	ⓑ	ⓒ	ⓓ	50.	ⓐ	ⓑ	ⓒ	ⓓ

PRETEST

1. Which of the following represents 5% as a decimal?
   a. .005
   b. .05
   c. 0.5
   d. 5

2. In the figure below, what percentage of the inventory is Walnut and Pine?

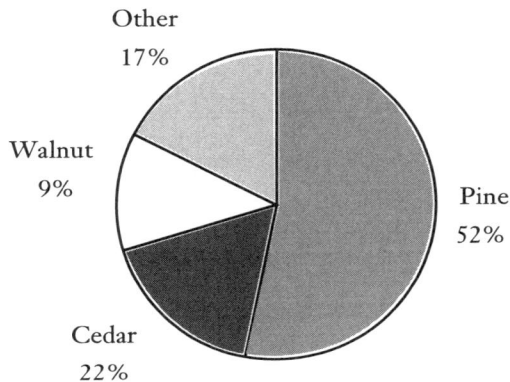

   a. 39%
   b. 44%
   c. 61%
   d. 78%

3. Guests in a hotel have been charged $168.50 for their stay. When the guests review their bill, they notice that there is an overcharge of $13.20. When the mistake is corrected, what will the new bill be?
   a. $142.10
   b. $155.30
   c. $168.50
   d. $181.70

4. A tanker truck is $\frac{4}{5}$ full, what is this fraction represented as a percent?
   a. 40%
   b. 45%
   c. 75%
   d. 80%

5. The window below must be fitted for window coverings. If each of the four panes is 8″ wide and 10″ tall, how large does the entire window covering need to be? Ignore the window frame in your calculation.

    a. 16″ wide and 20″ tall
    b. 20″ wide and 16″ tall
    c. 18″ wide and 24″ tall
    d. 32″ wide and 40″ tall

6. You have 274.8 inches of electrical wiring. How many feet of wiring do you have?
    a. 19.6 feet
    b. 22.9 feet
    c. 23.4 feet
    d. 27.48 feet

7. What is the approximate length of the nail below?

    a. 4.5″
    b. 5″
    c. 5.5″
    d. 9″

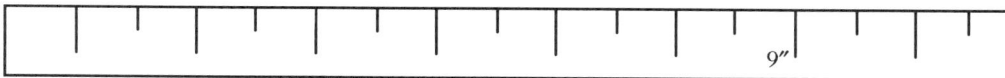

8. A cleaning service pays an employee $90 for a particular job. If the employee earns $15 per hour, how many hours did she work?
    a. 5 hours
    b. 5.5 hours
    c. 6 hours
    d. 10 hours

9. A landscape designer purchases 12 plants for $7.45 each. What is his total bill without tax?
   a. $84.40
   b. $89.40
   c. $90.35
   d. $92.35

10. A customer hands you $20 for a purchase total of $14.52. How much change is due back to the customer?
   a. $5.48
   b. $5.58
   c. $6.48
   d. $6.58

11. Ten bagels is what percentage of a dozen bagels?
   a. 16.7%
   b. 76.9%
   c. 81.4%
   d. 83.3%

12. The state sales tax is 7.15%. What is the total cost for a purchase that costs $20 pre-tax?
   a. $1.43
   b. $21.43
   c. $22.53
   d. 26.33

13. This morning, 160 heads of lettuce were delivered to a local grocer. Due to weather and traveling conditions, 30% of the produce was not suitable for sale. How many heads of lettuce were spoiled?
   a. 40
   b. 42
   c. 48
   d. 56

14. At the start of the day, inventory reports showed that there were 37 drills in stock at the store. After a Father's Day sale, the receipts showed that 9 drills were sold. How many drills were left in stock?
   a. 25
   b. 26
   c. 28
   d. 46

Use the figure below to answer questions 15–16.

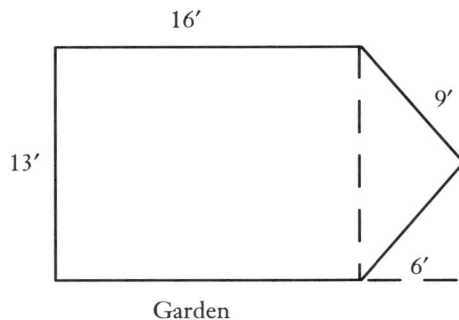

Garden

15. A landscaper needs to surround the garden with plastic tubing. How many feet of tubing are required?
  a. 38 feet
  b. 45 feet
  c. 58 feet
  d. 63 feet

16. Now the landscaper needs to cover the garden with mulch. In order to do so, he needs to calculate the area of the garden. What is that area of the garden?
  a. 234 ft^2
  b. 247 ft^2
  c. 262 ft^2
  d. not enough information

17. The manager of a hardware store is gathering scraps from boards that were originally 12 feet in length. Out of the five pieces of scrap, two were $\frac{1}{4}$ of the original length, one was $\frac{1}{3}$ of the original, one was $\frac{2}{5}$ the original, and the last one was $\frac{1}{2}$ the original. Find the length, in feet, of all the scraps.
  a. 20.8 feet
  b. 18.7 feet
  c. 17.8 feet
  d. 21 feet

18. For long distance calls a telephone company is charging a $0.32 connection fee and 8¢ a minute for calls out of state. How much does it cost to make a 1 hour and 15 minute call out of state?
  a. $0.42
  b. $4.20
  c. $6.00
  d. $6.32

19. A contractor received $250 upfront to finish a job. He needed to make several purchases to finish the job. The first supplies cost $135.60. However, the contractor returned $12.45 worth of supplies and purchased an additional $69.15 of supplies. How much money is left over?
   a. $32.80
   b. $45.25
   c. $57.70
   d. There is no money left over.

20. A chef is preparing for a large event and must prepare 88 servings of a recipe calling for 2.75 grams of duck per serving. The chef orders in kilograms, so how many kilograms of duck must he purchase?
   a. 0.242 kg
   b. 2.42 kg
   c. 24.2 kg
   d. 242 kg

21. Baseboards are sold in 16-linear-feet-sections. How many baseboards are necessary to complete a room with 152-linear-feet-of-walls?
   a. nine boards
   b. ten boards
   c. eleven boards
   d. twelve boards

22. At a fabric store a customer wants to purchase 5′ of a particular fabric that sells for $4.50 per yard. How much will it cost for a 5′ section?
   a. $7.50
   b. $8.35
   c. $21.25
   d. $22.50

23. Today, 200 copies of a new book arrived at the local bookstore. Unfortunately, 50 books were damaged. Of those 50 books, 13 could still be sold at a discounted price at an outlet store. What percentage of the damaged stock could still be sold?
   a. 26%
   b. 6%
   c. 25%
   d. 75%

24. A roll of carpet is 400 ft long. A carpet layer has $\frac{1}{4}$ of a roll left and the customer only needs $\frac{1}{2}$ of the leftover roll. How much length in carpet is needed?

   a. 50 ft

   b. 62 ft

   c. 120 ft

   d. 200 ft

25. A moving company is offering a 15% discount for customers during the week. If the weekend price quote for a family was $500, what is the new price with the discount?

   a. $75

   b. $400

   c. $425

   d. $450

26. After a basement flooding, a plumber was called in to fix the problem. The plumber worked from 9:00 A.M. until 4:00 P.M. with a one-hour, unpaid lunch break. If he charges $25 per hour, how much will the bill for his services be?

   a. $150

   b. $175

   c. $200

   d. $225

Use the figure below to answer questions 27–29.

Production

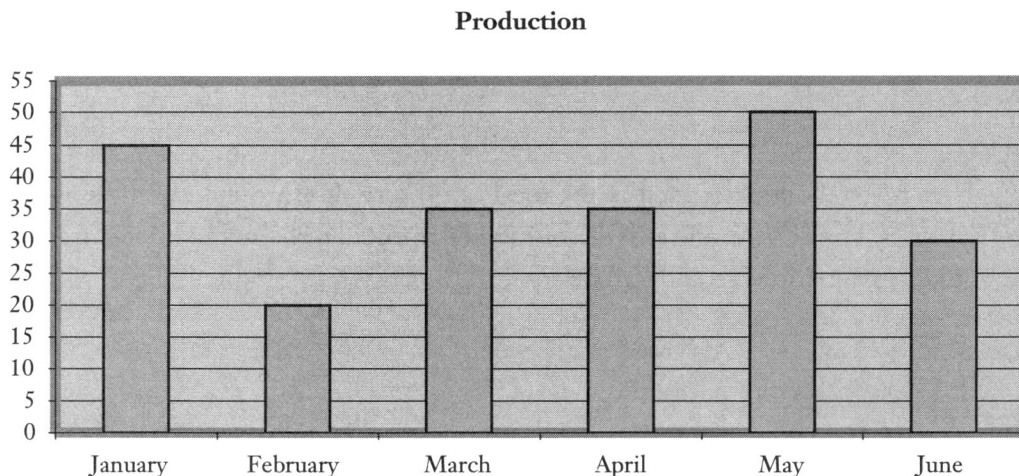

27. A local merchant charted unit production for the first six months of the year. According to the chart, what fraction of the units were produced in April?

    a. $\frac{1}{215}$

    b. 0.162

    c. $\frac{7}{43}$

    d. $\frac{3}{20}$

28. According to the chart, what is the average monthly production?

    a. 32.5 units per month

    b. 35.8 units per month

    c. 35 units per month

    d. 36.2 units per month

29. According to the chart, what was the increase in unit production from February to March?

    a. 45.2%

    b. 57.1%

    c. 75%

    d. 80%

30. A painter is commissioned to repaint the four walls in each of a company's five offices. The dimensions of the walls are approximately 11×8 feet with no windows. If one gallon of paint can cover 500 ft^2 of wall space, how many gallons are needed to complete the job?

    a. two gallons

    b. three gallons

    c. four gallons

    d. five gallons

31. Paint is sold at $24 for one gallon or $90 for five gallons. The interior of a new construction home requires 19 gallons of paint. What is the most cost efficient way to purchase the paint?

    a. 19 one-gallon containers

    b. 4 five-gallon containers

    c. 3 five-gallon containers and 4 one-gallon containers

    d. 2 five-gallon containers and 9 one-gallon containers

32. An oil tanker started the day $\frac{2}{3}$ full before making a stop that depleted the supply by $\frac{1}{2}$. After another stop, the tanker lost $\frac{3}{4}$ more of the remaining oil. If the tanker can hold a total of 255 gallons, how much was left at the end of the day?

    a. Impossible, the tanker ran out of oil before the end.

    b. 10.63 gallons

    c. 21.25 gallons

    d. 63.75 gallons

33. A cylindrical silo has the capacity to hold 8,478 ft³ of grain more than it is currently holding. At 25% capacity, the grain measures a height of 9′ and contains 2,826 ft³ of the supply. What is the diameter of the silo? Use 3.14 for π.

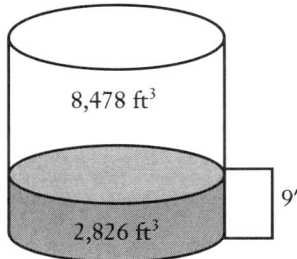

a. 10 feet
b. 13 feet
c. 20 feet
d. 22 feet

34. An Olympic size swimming pool has a leak that drains 7 gallons of water per day. If the pool system automatically cycles in $1\frac{1}{2}$ gallons every 10 hours, what will the water level be after 60 days?
a. there is no difference
b. 420 gallons lower
c. 204 gallons lower
d. 318 gallons lower

35. A carpet layer needs to carpet a rectangular bedroom. He charges a labor fee of $3.50 per square foot, plus materials, but because he is friends with the client, he is offering a 7% discount on the total bill. If the length of the room is 10 ft and the width is 12 ft, and the carpet costs $12.94 per square foot, approximately how much will the carpet layer charge after the discount?
a. $1,972.80
b. $138.10
c. $1,834.70
d. $1,552.80

36. A bookstore owner buys a particular book at a price of $3.65 per book. She purchased 350 copies of it and needs to earn a profit of at least $1,500. Assuming she is guaranteed to sell at least 65% of the stock, how much should she charge per book, keeping in mind she must keep competitive prices?
a. $6.58
b. $9.78
c. $12.19
d. $17.50

37. A bookshelf was originally sold for $125. Before a big sale, the price was increased 20% and then discounted 30%. What is the selling price now?

   a. $105.00
   b. $109.00
   c. $112.50
   d. $112.75

38. Due to continuous expansion, maintenance, and increased membership, a local club's maintenance fees are on the rise. Below is a graph charting the costs of the last few months. Assuming that the trend continues, how much money should the manager expect to have to pay in October?

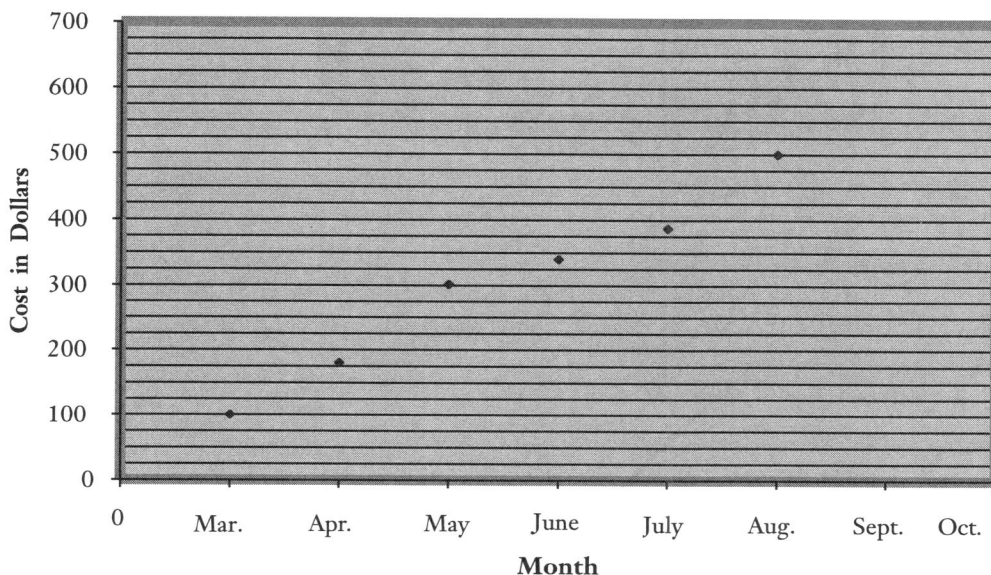

   a. There is no way to tell.
   b. around $600
   c. around $635
   d. around $685

39. A contractor has 80 feet of fencing to enclose a Jacuzzi area. There should be a walkway around the Jacuzzi that is 3 feet wide. What shape of Jacuzzi should be built in order to have the largest amount of area with the given perimeter?

   a. a square
   b. a cross
   c. a rectangle
   d. a circle

40. The chart below shows the purchasing trends of customers in a given year; 320 customers exclusively purchase brand A, 120 customers exclusively purchase brand B, and 40 customers purchase either one differing year to year. What percentage of customers typically purchase brand B during any given month?

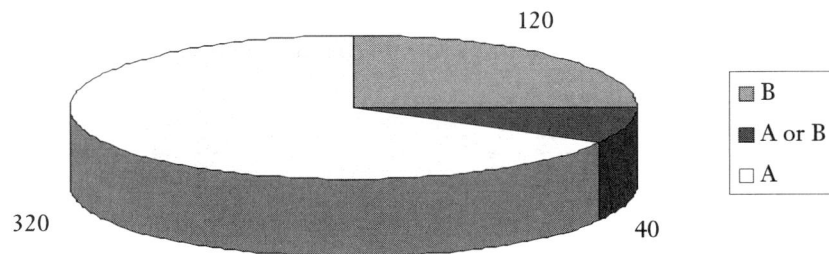

a. 25%
b. up to 33.3%
c. 68.4%
d. cannot be determined

41. The graph represents a car's speed and the miles per gallon calculated when driving a particular speed. What range of speeds guarantees at least a 40 miles per gallon result?

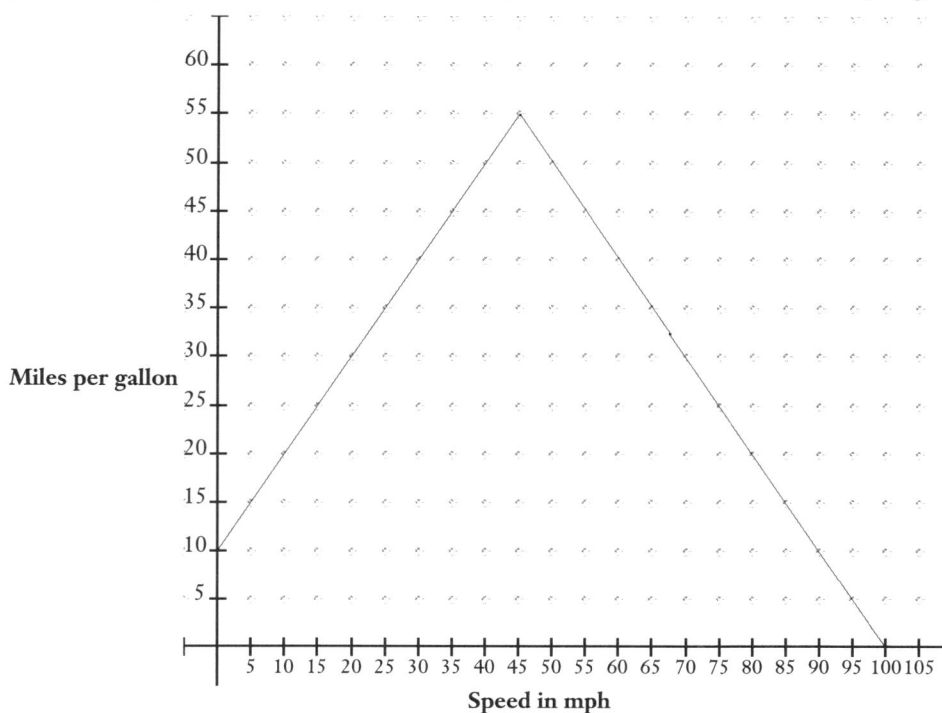

a. 30–60 mph

b. 45–90 mph

c. 35–55 mph

d. not enough information

42. The height of each pane in the window below is $\frac{2}{3}$ of the width of each pane. The area of one pane is 216 in^2, what are the dimensions of the entire window?

a. $4' \times 4.5'$

b. $18'' \times 12''$

c. $18' \times 27'$

d. $48' \times 54'$

43. In a large corporation, the CEO has decided to set-up the management team so that each employee has two bosses above him/her. If there are six employees at the lowest positions, how many employees are there in the first five levels of management? (The diagram shows a partial look to get a better understanding.)

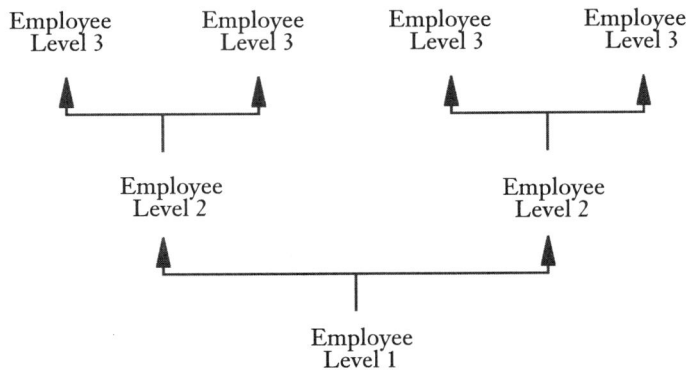

a. 32 employees
b. 36 employees
c. 148 employees
d. 186 employees

44. Todd gets time and a half for hours worked over his required 40 hours per week. What is his regular hourly wage if he made a total of $725 for a 52-hour workweek?
a. $12.10/hour
b. $12.50/hour
c. $18.13/hour
d. $18.75/hour

45. A technical drawing has been created to show the dimensions of a basement. The width of the walkway from Room B–2 measures 1.125″ and the width of Room B–1 is 3″. If the scale is 1″: 4′, what is the real area of the basement?

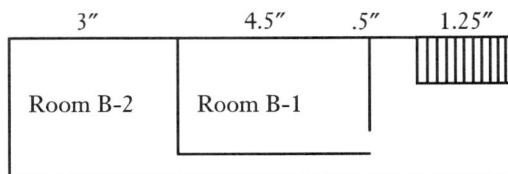

a. 152.6 ft²
b. 38.16 ft²
c. 610.5 ft²
d. 1,124.4 ft²

46. A cube-shaped piece of metal must be coated with a special Teflon covering. If the cube has a side of 15 cm, what is the total surface area that needs to be coated?
   a. 600 cm²
   b. 900 cm²
   c. 1350 cm²
   d. 3375 cm²

47. During construction, an $8\frac{1}{2}$ ft wall needs a support beam running from the top of the wall down to the ground about $10\frac{1}{2}$ ft away from the base of the wall. How long should the support beam be?
   a. 10.5 feet
   b. 13.5 feet
   c. 16.8 feet
   d. 20 feet

48. A business owner has a credit card for the company. The credit card charges 8.9% interest for any balance that is not paid in full each month. How much money will the company owe in two months if there is currently a $4,535 balance on the card and there are no additional purchases?
   a. $843.16
   b. $4,216.16
   c. $5,378.16
   d. $6,112.16

49. Camille and Katie are working together to tile the bottom floor of house. If Camille could finish the job alone in 16.5 hours and Katie could finish it alone in 22.25 hours, how long will it take them to finish the job together?
   a. 9.47 hours
   b. 10.42 hours
   c. 11.94 hours
   d. 19.38 hours

50. Delivery-R-Us charges a $25 flat fee and $15.50 for each package delivered. Deliver–2-U charges a flat fee of $10 and $16.75 for each package. Which company offers the best deal?
   a. Delivery-R-Us
   b. Deliver–2-U
   c. Delivery-R-Us until you have 5 packages
   d. Deliver–2-U until you have 12 packages

ANSWERS

1. b. Percentages are numbers divided by 100 ($\frac{5}{100}$ = .05). You can also move the decimal point 2 places to the left.

$$0.05$$

2. c. Take the percentage of Pine and add it to the percentage of Walnut for the combined percentage (52 + 9 = 61); 61% of the wood is Walnut and Pine.

3. b. The guests are refunded $13.20, which means it must be subtracted from their bill (168.50 − 13.20 = 155.30); $155.30 is the new charge.

4. d. The first step is to find out what $\frac{4}{5}$ is as a decimal ($\frac{4}{5}$ = .8). A decimal can be converted to a fraction by multiplying it by 100 (.8 × 100 = 80); 80%.

5. a. There are two panes across the top for a total of 16″ across (8 + 8 = 16). There are two panes top to bottom on one side for a total of 20″ high (10 + 10 = 20); 16″ wide and 20″ tall.

6. b. To convert inches into feet you must divide by 12 because there are 12 inches for every foot ($\frac{274.8}{12}$ = 22.9); you have 22.9 feet.

7. a. You can count in intervals, or you can use subtraction (9 − 4.5 = 4.5) to determine the length of the nail; 4.5″.

8. c. In order to find out how many hours she worked, divide the total amount paid by the hourly wage ($\frac{90}{15}$ = 6); 6 hours.

9. b. When purchasing multiple items for the same price, use multiplication to find the answer (12 × 7.45 = 89.40); $89.40.

10. a. Subtract the total purchase from the amount the customer gave you (20 − 14.52 = 5.48); $5.48 is the change.

11. d. There are 12 bagels in a dozen bagels, therefore 10 out of 12 bagels is .833 ($\frac{10}{12}$ = .833) and to convert into a percentage, multiply by 100 (.833 × 100 = 83.3%).

12. b. The tax for this purchase is calculated by multiplying the pre-tax price by the percent of tax written in decimal form; (20 × 0.0715 = 1.43). Then, add the tax to the original sales amount for the total (20 + 1.43 = 21.43); $21.43 is the total.

13. c. The decimal .30 represents 30%. Multiply the percent damaged by the number of lettuce heads delivered, 0.30 × 160 = 48; 48 heads of lettuce were spoiled.

14. c. Take the original amount and subtract away the sold items (37 − 9 = 28); 28 drills are left.

15. d. Perimeter is calculated by adding the length of all the sides (13 + 16 + 9 + 9 + 16 = 63); 63 feet are necessary.

16. b. Split up the garden into the rectangle and the triangle and add up their separate areas. The area of the rectangle is length × width (16 × 13 = 208) and the area of the triangle is $\frac{1}{2}$ base × height ($\frac{1}{2}$ × 6 × 13 = 39). Add the two together to calculate the total (208 + 39 = 247); 247 ft².

17. a. In order to find the total amount, you must add up the length all of the pieces ($\frac{1}{4}$ + $\frac{1}{4}$ + $\frac{1}{3}$ + $\frac{2}{5}$ + $\frac{1}{2}$). The first two pieces are $\frac{1}{4}$ of 12 feet, or 3 feet and 3 feet. The next piece is $\frac{1}{3}$ of 12 feet, or 4 feet. The fourth piece is $\frac{2}{5}$ of 12 feet, or 4.8 feet. The final piece is $\frac{1}{2}$ of 12 feet, or 6 feet. Now you must add the lengths together to find the total length. Therefore, 3 + 3 + 4 + 4.8 + 6 = 20.8 feet.

18. **d.** Start by converting hours to minutes, thus, 1 hour and 15 minutes is equivalent to 75 minutes (1 hour = 60 minutes + 15 minutes = 75 minutes) and 75 minutes with $.08/minute costs $6.00 (75 × .08 = 6). The last step is to add on the connection fee of $0.32 for a total of $6.32.

19. **c.** Take the original amount and subtract the expenses (250 − 135.60 − 69.15 = 45.25) and then add the returned amount because it was refunded (45.25 + 12.45 = 57.70) to find the total left, $57.70.

20. **a.** For 88 plates, 242 grams are needed (88 × 2.75 = 242). The chef must order in kg; in order to convert to kg, divide the grams by 1,000 ($\frac{242}{1,000}$ = .242); 0.242 kg are needed.

21. **b.** Ten boards cover 160 linear feet (16 × 10 = 160) and nine boards only cover 144 linear feet. You must buy enough baseboard to cover the entire length.

22. **a.** The material is sold in yards but the purchase is in feet. Convert feet into yards by dividing by 3 (3 feet in 1 yard) and then multiply by the cost per yard ($\frac{5}{3}$ = 1.67, 1.67 × 4.50 = 7.5); $7.50 is the cost for 5′ of material.

23. **a.** 50 of the 200 books were damaged, however, 13 of the 50 damaged books could still be sold ($\frac{13}{50}$ = .26 or 26%.). Be careful not to use the original number of 200 to calculate the percentage.

24. **a.** Originally there was 400 ft and $\frac{1}{4}$ of that is 100 ft (400 × $\frac{1}{4}$ = 100) and $\frac{1}{2}$ of that is 50 feet (100 × $\frac{1}{2}$ = 50); 50 ft of carpet is needed.

25. **c.** The discounted price is $425. The discount is $75 (.15 × 500 = 75) and that needs to be subtracted from the original price (500 − 75 = 425). You can also solve this by taking the percentage that will be paid, 85% (1.00 − .15 = .85) and multiplying that by the original price (.85 × 500 = 425).

26. **a.** The plumber worked for 6 hours (7 total but he took a one-hour lunch break) and he charges $25/hour for a total of $150 (6 × 25 = 150); $150 will be charged.

27. **c.** First you must find the total production for the six months (45 + 20 + 35 + 35 + 50 + 30 = 215). April produced 35 units, which can be made into a fraction by placing 35 over the total units (215). This fraction can be reduced by dividing the top and bottom by the lowest common factor, which in this case is 5; ($\frac{35}{215}$ = $\frac{7}{43}$); $\frac{7}{43}$ is the total production.

28. **b.** To find the average, add up the total and divide by the number of months ($\frac{215}{6}$ = 35.8); 35.8 per month.

29. **c.** February's production was 20 and March's production was 35. The difference in production was 15 (35 − 20 = 15) and 15 of the original 20 is represented by 75% ($\frac{15}{20}$); 75%.

30. **c.** The area of a rectangle is length × width, so each office has approximately 352 ft² of wall space (8 × 11 = 88, and there are four walls per office, 88 × 4 = 352). For five offices there are 1,760 ft² of walls (352 × 5 = 1,760) and three gallons only covers 1,500 ft² (500 × 3 = 1,500) so the painter will need one more gallon; four gallons are needed.

31. **b.** Four of the 5-gallon containers will provide 20 gallons (4 × 5 = 20), for a price of $360 (4 × 90 = 360). Purchasing three 5-gallon containers costs $270 (3 × 90 = 270) and only covers 15 gallons (3 × 5 = 15), so four more single gallons are needed (19 – 15 = 4), which costs $96 for a total of $366 (270 + 96 = 366)—$6 more than purchasing one gallon too much.

32. **c.** There were 21.25 gallons remaining at the end of the day. The tanker started at 170 gallons (255 × $\frac{2}{3}$ = 170) and lost 85 gallons at the first start (170 × $\frac{1}{2}$ = 85), so 85 gallons is left (170 – 85 = 85). The next stop lost 63.75 gallons (85 × $\frac{3}{4}$ = 63.75) for a total of 21.25 gallons (85 – 63.75 = 21.25).

33. **c.** The diameter is 20 feet. The silo can hold a total of 11,304 ft3 of grain (2,826 + 8,478 = 11,304) and the total height is 36 ft (9 × 4 = 36). Volume of a cylinder is $V = \pi r^2 h$ so that r^2 must be 100 (11,304 = 3.14 × r^2 × 36, $r^2 = \frac{11,304}{113.04} = 100$) and the square root of 100 is 10; r stands for the radius, which is $\frac{1}{2}$ of the diameter, so double the radius to get 20 feet for the diameter (10 × 2 = 20).

34. **c.** 204 gallons lower. The pool is draining 7 gal/day and the pool is only refilling 3.6 gal/day ($\frac{1.5\ gal}{10\ hours} \times \frac{24\ hours}{1\ day} = \frac{1.5 \times 24}{10} = 3.6$). The pool is losing only 3.4 gal/day (7 – 3.6 = 3.4) so after 60 days it will lose 204 gallons (3.4 × 60 = 204).

35. **c.** First, find the area of the room using the formula $A = lw$, where l is the length of the room and w is the width. Substitute into the formula to get $A = (10)(12) = 120$ ft². Start by multiplying the total area by $12.94 to get the cost of the carpet; 120 × 12.94 = $1,552.80. To calculate the labor cost, multiply the total area by the rate of $3.50 per square foot (120 × 3.5 = $420). Now you have to add the two numbers to calculate the cost of labor and materials; (420 + 1,552.8 = $1,972.80). Remember, for this job there is a 7% discount. Find 7% (.07) of $1,972.80 (.07 × $1,972.80 = 138.096 or 138.10) and subtract that number from the total fee ($1,972.80 – 138.10 = 1,834.70).

36. **c.** The question states that the bookstore owner is guaranteed to sell at least 228 books (350 × .65 = 227.5 books, round up to 228). We also know she paid $1,277.50 for purchasing the books from the supplier (350 × 3.65 = 1,277.50). But, she needs to sell enough books not only to break even ($1,277.50), but also to make a profit of $1,500. That means she must sell enough books to make $2,777.50. Split that total (2,777.50) by the number of books that are guaranteed to sell. ($\frac{2,777.50}{228} = 12.182$). This means she must sell the books for at least $12.19 each. If she chooses to sell each book at $17.50 per book, she risks losing sales due to high prices.

37. **a.** $105.00 is the new sale price. The increase before the sale brought up the price to $150 (120% of the original price: 1.20 × 125 = 150) and then a 30% reduction took the price down to $105 (70% of the new sales price: .70 × 150 = 105).

38. **c.** The manager should expect to pay around $635 for October. The data provides a linear relationship, which means that the costs are increasing at a steady rate. The fastest way to get a good approximation is to draw a line through the data, trying to get as many data points above the line as there are below. Then, match up the line with October and read off the cost-axis.

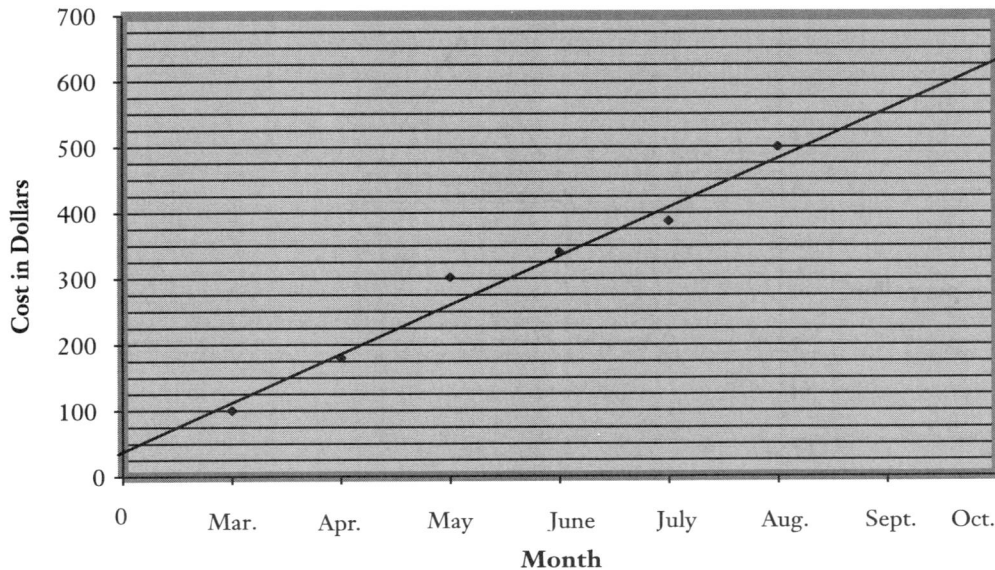

39. **d.** A circle will provide the largest area. In order to find the area of the circle, you must find the radius using the circumference (perimeter), which is 12.73 ft ($C = 3.14 \times 2 \times r$; $80 = 3.14 \times 2 \times r$; $r = 12.73$). The area of a circle is calculated by $A = 3.14r^2$ giving a total of 509.8 ft². If a square shape were built, the sides would be 20×20 ($\frac{80}{4} = 20$) giving an area of only 400 ft² ($20 \times 20 = 400$). Playing around with numbers for a rectangle will show that the area would not be able to exceed the circle.

40. **b.** If the purchases are spread out pretty evenly throughout the year then it is possible for up to 33.3% of the sales to be for brand B. There are a total of 480 customers (120 + 40 + 320) and at most for the year there are 160 customers for brand B (120 + 40 = 160). This is represented by 33.3% ($\frac{160}{480} = .333$).

41. a. When driving anywhere from 30–60 mph the car will return at least a 40 miles per gallon statistic. You need to find the values on the horizontal axis that will make the vertical answers 40 or larger. If the car exceeds 60 mph, efficiency is actually lost and the mpg drops below 40.

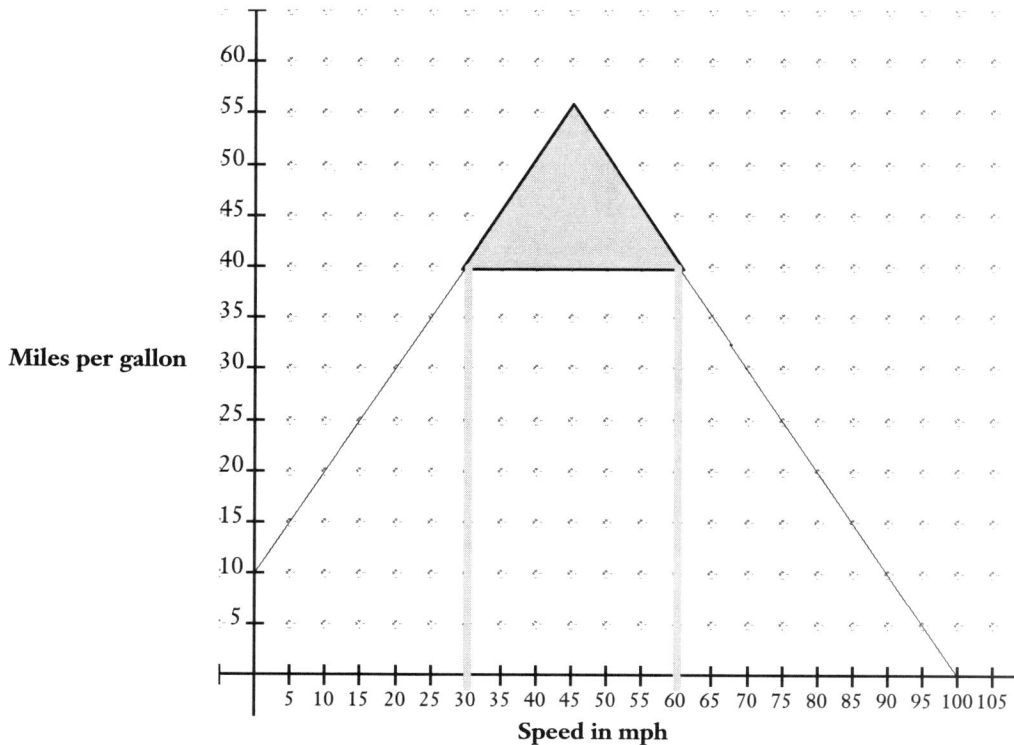

Miles per gallon

Speed in mph

42. a. For this question, remember that 12 inches = 1 foot. The dimensions of the entire window are $4' \times 4.5'$. One individual pane is 216 ft^2 and looking at the diagram to the side, $\frac{2}{3}x \bullet x = 216$, so the width of one pane is 18" ($216 \times \frac{3}{2}$, $x^2 = 324$, the square root of 324 is 18). If the length is 18 then the height of the pane is 12 ($18 \times \frac{2}{3} = 12$). There are three panes across, so the width of the entire window is 54" ($18 \times 3 = 54$) and there are four panes vertically, so the window is 48" tall ($12 \times 4 = 48$); 54" is equivalent to 4.5' ($\frac{54}{12} = 4.5$), and 48" is equivalent to 4' ($\frac{48}{12}$).

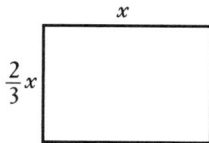

43. **d.** There are a total of 186 employees in the first five levels. This is an exponential problem; each employee has two bosses directly ahead. In the first level we know that we have six employees. The number of bosses in the level ahead is 2^1 for each one giving a total of 12 (see table below).

LEVEL	# OF EMPLOYEES
1	= 6
2	$= 6(2^1) = 6 \cdot 2 = 12$
3	$= 6(2^2) = 6 \cdot 4 = 24$
4	$= 6(2^3) = 6 \cdot 8 = 48$
5	$= 6(2^4) = 6 \cdot 16 = 96$
Total	$= 6 + 12 + 24 + 48 + 96 = 186$

44. **b.** Todd's regular rate is x and his overtime rate is $1.5x$. He worked a total of 52 hours, so he worked his regular 40 hours as well as 12 hours of overtime ($52 - 40 = 12$). The money he earned from the regular week is time × rate, which is represented by $40x$ ($40 \times x = 40x$). The money he earned from overtime work is also time × rate, which is $18x$ ($12 \times 1.5x = 18x$). His regular salary plus overtime gives the total and can be written as $40x + 18x = 725$. Solve for x:

$$58x = 725$$
$$\frac{58x}{58} = \frac{725}{58}$$
$$x = 12.5$$

So Todd makes $12.50 per hour.

45. **c.** The real area of the basement is 610.5 ft². In order to find the area, the dimensions of the room must be calculated. Add up all the lengths for the top and bottom for a total of 9.25″ (3 + 4.5 + .5 + 1.25 = 9.25) and the sides for a total of 4.125″ (3 + 1.125 = 4.125). Before you can calculate the area, the dimensions must be converted into feet; 9.25″ = 37′ (9.25 × 4 = 37) and 4.125″ = 16.5′ (4.125 × 4 = 16.5). Finally, to find the area, multiply the length times the width for a total of 610.5 ft² (37 × 16.5 = 610.5).

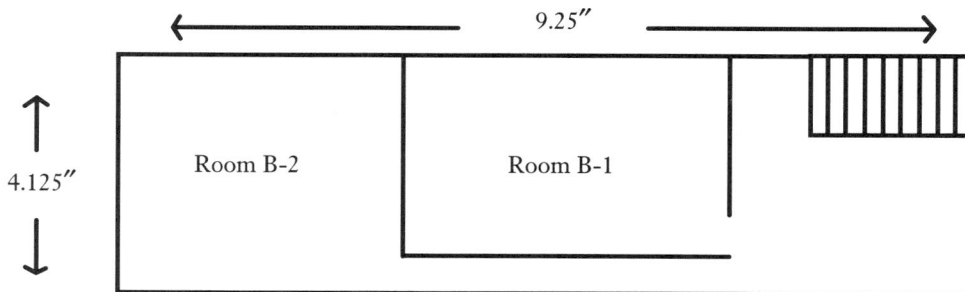

46. **c.** The total surface area is 1,350 cm². Find the surface area of one side and then multiply by six because a cube has six equal faces. The surface area of one side is 225 (15² = 15 × 15 = 225), for a total of 1,350 (6 × 225 = 1,350).

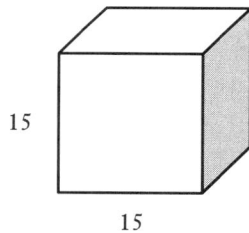

47. **b.** The support should be approximately 13.5 feet long.

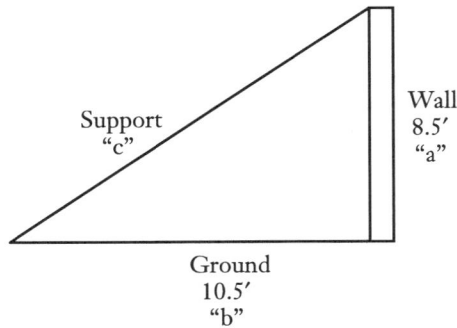

The guide forms a right triangle with the wall as seen in the diagram. Using the Pythagorean theorem, solve for c.

$$a^2 + b^2 = c^2 \rightarrow (10.5)^2 + (8.5)^2 = c^2$$
$$110.25 + 72.25 = c^2 \rightarrow 182.5 = c^2$$
$$\sqrt{182.5} = 13.5$$

48. **c.** After the first month, the company will owe \$403.62 for interest ($4{,}535 \times .089 = 403.615$) and that will be added to the balance ($4{,}535 + 403.62 = 4{,}938.62$) to calculate what is due the next month. The second month's interest is 439.54 ($4{,}938.62 \times .089 = 439.537$) for a total balance of \$5,378.16 ($4{,}938.62 + 439.54 = 5{,}378.16$) after two months.

49. **a.** This is a work problem where Camille and Katie are working together to finish a project. Camille's rate is $\frac{1}{16.5}$, Katie's rate is $\frac{1}{22.25}$, and the rate together is $\frac{1}{t}$. If you take the ladies' rates and add them together, you will find the total rate.

$$\frac{1}{16.5} + \frac{1}{22.25} = \frac{1}{t}$$
$$\left(\frac{1}{16.5} + \frac{1}{22.25} = \frac{1}{t}\right) \cdot t(16.5)(22.25)$$
$$22.25t + 16.5t = (16.5)(22.25)$$
$$\frac{38.75t}{38.75} = \frac{367.125}{38.75}$$
$$t = 9.474$$

50. **d.** Deliver–2-U is a better deal until you need to send more than 12 packages. You can solve this using a chart, equation, or graph.

# PKGS	DELIVERY-R-US	DELIVER-2-U
1	25 + 15.50= 40.50	10 + 16.75 = 26.75
5	25 + 5(15.50) = 102.50	10 + 5(16.75) = 93.75
10	25 + 10(15.50) = 180.00	10 + 10(16.75) = 177.50
15	25 + 15(15.50) = 257.50	10 + 15(16.75) = 261.25
12	25 + 12(15.50) = 211.00	10 + 12(16.75) = 211.00

You can generate two linear equations from the given data. Delivery-R-Us is *cost* = 15.50(*p*) + 25 and Deliver–2-U is *cost* = 16.75(*p*) + 10 where *p* represents the number of packages purchased. Similar to the chart, plug in values for *p* to find out what happens to the cost.

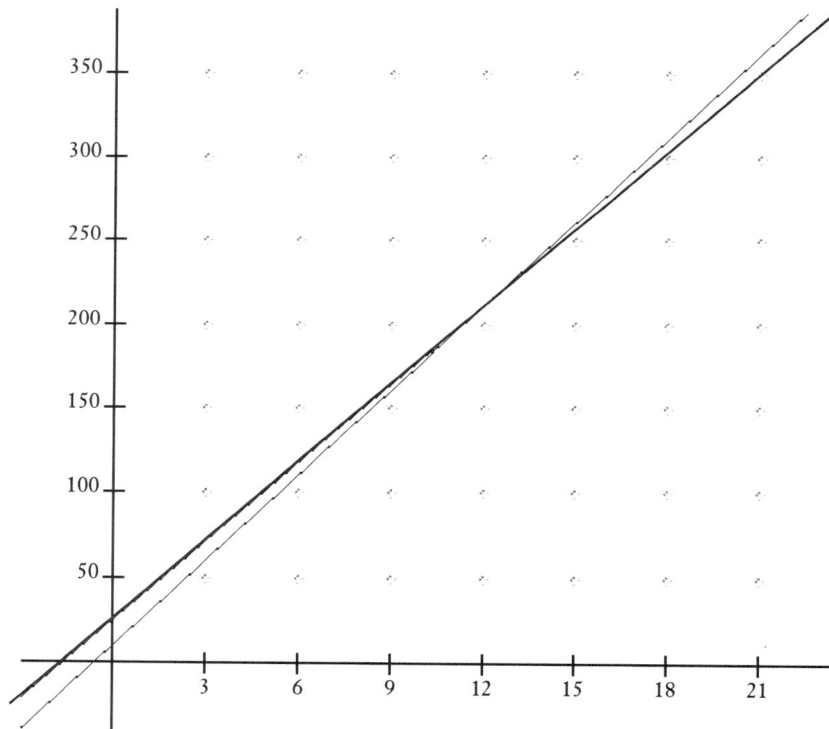

If the two equations are graphed, it is easy to see how the cost changes. The line that is lower represents the smaller cross, they meet at one point where the cost is the same and then the other company is less expensive. The darker line represents Delivery-R-Us.

3

Arithmetic Review

This chapter covers the basics of mathematical operations and their sequence. It also reviews variables, integers, fractions, decimals, and square roots.

Basic problem solving in mathematics is rooted in whole number math facts, mainly addition facts and multiplication tables. If you are unsure of any of these facts, now is the time to review. Make sure to memorize any parts of this review that you find troublesome. Your ability to work with numbers depends on how quickly and accurately you can do simple mathematical computations.

▶ NUMBERS AND SYMBOLS

Numbers
- ▶ **Whole numbers** include the counting numbers and zero: 0, 1, 2, 3, 4, 5, 6, . . .
- ▶ **Integers** include the whole numbers and their opposites. Remember, the opposite of zero is zero: . . . –3, –2, –1, 0, 1, 2, 3, . . .

▶ **Rational numbers** are all numbers that can be written as fractions, where the numerator and denominator are both integers, but the denominator is not zero. For example, $\frac{2}{3}$ is a rational number, as is $\frac{6}{5}$. The decimal form of these numbers is either a terminating (ending) decimal, such as the decimal form of $\frac{3}{4}$ which is 0.75; or a repeating decimal, such as the decimal form of $\frac{1}{3}$ which is 0.3333333. . . .

▶ **Irrational numbers** are numbers that cannot be expressed as terminating or repeating decimals (i.e. non-repeating, non-terminating decimals such as π, $\sqrt{2}$, $\sqrt{12}$).

Comparison Symbols

The following table will illustrate some comparison symbols.

=	is equal to	5 = 5
≠	is not equal to	4 ≠ 3
>	is greater than	5 > 3
≥	is greater than or equal to	$x \geq 5$ (*x* can be 5 or any number > 5)
<	is less than	4 < 6
≤	is less than or equal to	$x \leq 3$ (*x* can be 3 or any number < 3)

Symbols of Addition

In addition, the numbers being added are called **addends**. The result is called a **sum**. The symbol for addition is called a **plus** sign. In the following example, 4 and 5 are addends and 9 is the sum:

$$4 + 5 = 9$$

Symbols of Subtraction

In subtraction, the number being subtracted is called the **subtrahend**. The number being subtracted FROM is called the **minuend**. The answer to a subtraction problem is called a **difference**. The symbol for subtraction is called a **minus** sign. In the following example, 15 is the minuend, 4 is the subtrahend, and 11 is the difference.

$$15 - 4 = 11$$

Symbols of Multiplication

When two or more numbers are being multiplied, they are called **factors**. The answer that results is called the **product**. In the following example, 5 and 6 are factors and 30 is their product.

$$5 \times 6 = 30$$

There are several ways to represent multiplication in the previous mathematical statement.

▶ A dot between factors indicates multiplication.
5 • 6 = 30
▶ Parentheses around any one or more factors indicate multiplication.
(5)6 = 30, 5(6) = 30, and (5)(6) = 30
▶ Multiplication is also indicated when a number is placed next to a variable:
5a = 30. In this equation, 5 is being multiplied by a.

Symbols of Division

In division, the number being divided BY is called the **divisor**. The number being divided INTO is called the **dividend**. The answer to a division problem is called the **quotient**.

There are a few different ways to represent division with symbols. In each of the following equivalent expressions, 3 is the divisor and 8 is the dividend.

$$8 \div 3, \; 8/3, \; \frac{8}{3}, \; \text{or } 3\overline{)8}$$

Prime and Composite Numbers

A positive integer that is greater than the number 1 is either prime or composite, but not both.

▶ A prime number is a number that has exactly two factors.
Examples:
2, 3, 5, 7, 11, 13, 17, 19, 23 . . .
▶ A composite number is a number that has more than two factors.
Examples:
4, 6, 8, 9, 10, 12, 14, 15, 16 . . .
▶ The number 1 is neither prime nor composite since it has only one factor.

▶ OPERATIONS

Addition

Addition is used when it is necessary to combine amounts. It is easiest to add when the addends are stacked in a column with the place values aligned. Work from right to left, starting with the ones column.

Example:
Add 40 + 129 + 24

1. Align the addends in the ones column. Since it is necessary to work from right to left, begin to add starting with the ones column. Since the ones column totals 13, and 13 equals 1 ten

and 3 ones, write the 3 in the ones column of the answer, and regroup or "carry" the 1 ten to the next column as a 1 over the tens column so it gets added with the other tens:

```
  1
  40
 129
+ 24
   3
```

2. Add the tens column, including the regrouped 1.

```
  1
  40
 129
+ 24
  93
```

3. Then add the hundreds column. Since there is only one value, write the 1 in the answer.

```
  1
  40
 129
+ 24
 193
```

Subtraction

Subtraction is used to find the difference between amounts. It is easiest to subtract when the minuend and subtrahend are in a column with the place values aligned. Again, just as in addition, work from right to left. It may be necessary to regroup.

Example:
If Becky has 52 clients, and Claire has 36, how many more clients does Becky have?

1. Find the difference between their client numbers by subtracting. Start with the ones column. Since 2 is less than the number being subtracted (6), regroup or "borrow" a ten from the tens column. Add the regrouped amount to the ones column. Now subtract 12 – 6 in the ones column.

$$\begin{array}{r} {\overset{4}{\cancel{5}}}{\overset{1}{}}2 \\ -\ 36 \\ \hline 6 \end{array}$$

2. Regrouping 1 ten from the tens column left 4 tens. Subtract 4 – 3 and write the result in the tens column of the answer. Becky has 16 more clients than Claire. Check: 16 + 36 = 52.

$$\begin{array}{r} \overset{4}{\cancel{5}}\overset{1}{2} \\ -36 \\ \hline 16 \end{array}$$

Multiplication

In multiplication, the same amount is combined multiple times. For example, instead of adding 30 three times, 30 + 30 + 30, it is easier to simply multiply 30 by 3. If a problem asks for the product of two or more numbers, the numbers should be multiplied to arrive at the answer.

Example:

Find the product of 34 and 54.

1. Line up the place values vertically, writing the problem in columns. Multiply the number in the ones place of the top factor (4) by the number in the ones place of the bottom factor (4): $4 \times 4 = 16$. Since 16 = 1 ten and 6 ones, write the 6 in the ones place in the first partial product. Regroup or "carry" the ten by writing a 1 above the tens place of the top factor.

$$\begin{array}{r} \overset{1}{} \\ 34 \\ \times\ 54 \\ \hline 6 \end{array}$$

2. Multiply the number in the tens place in the top factor (3) by the number in the ones place of the bottom factor (4); $4 \times 3 = 12$. Then add the regrouped amount 12 + 1 = 13. Write the 3 in the tens column and the 1 in the hundreds column of the partial product.

$$\begin{array}{r} \overset{1}{} \\ 34 \\ \times\ 54 \\ \hline 136 \end{array}$$

3. The last calculations to be done require multiplying by the tens place of the bottom factor. Multiply 5 (tens from bottom factor) by 4 (ones from top factor); $5 \times 4 = 20$, but since the 5 really represents a number of tens, the actual value of the answer is 200 ($50 \times 4 = 200$). Therefore, write the two zeros under the ones and tens columns of the second partial product and regroup or "carry" the 2 hundreds by writing a 2 above the tens place of the top factor.

$$\begin{array}{r} \overset{2}{} \\ 34 \\ \times\ 54 \\ \hline 6 \\ 00 \end{array}$$

4. Multiply 5 (tens from bottom factor) by 3 (tens from top factor); $5 \times 3 = 15$, but since the 5 and the 3 each represent a number of tens, the actual value of the answer is 1,500 ($50 \times 30 = 1,500$). Add the two additional hundreds carried over from the last multiplication: $15 + 2 = 17$ (hundreds). Write the 17 in front of the zeros in the second partial product.

$$
\begin{array}{r}
{\scriptstyle 2} \\
34 \\
\times\ 54 \\
\hline
136 \\
1,700 \\
\end{array}
$$

5. Add the partial products to find the total product:

$$
\begin{array}{r}
{\scriptstyle 2} \\
34 \\
\times\ 54 \\
\hline
136 \\
+1,700 \\
\hline
1,836 \\
\end{array}
$$

Division

In division, the same amount is subtracted multiple times. For example, instead of subtracting 5 from 25 as many times as possible, $25 - 5 - 5 - 5 - 5 - 5$, it is easier to simply divide, asking how many 5s are in 25?: $25 \div 5$.

Example:

At a road show, three artists sold their beads for a total of $54. If they share the money equally, how much money should each artist receive?

1. Divide the total amount ($54) by the number of ways the money is to be split (3). Work from left to right. How many times does 3 divide 5? Write the answer, 1, directly above the 5 in the dividend, since both the 5 and the 1 represent a number of tens. Now multiply: since $1(\text{ten}) \times 3(\text{ones}) = 3(\text{tens})$, write the 3 under the 5, and subtract. $5(\text{tens}) - 3(\text{tens}) = 2(\text{tens})$.

$$
\begin{array}{r}
1 \\
3\overline{)54} \\
-3 \\
\hline
2 \\
\end{array}
$$

2. Continue dividing. Bring down the 4 from the ones place in the dividend. How many times does 3 divide 24? Write the answer, 8, directly above the 4 in the dividend. Since $3 \times 8 = 24$, write 24 below the other 24 and subtract $24 - 24 = 0$.

$$
\begin{array}{r}
1 \\
3\overline{)54} \\
-3\!\downarrow \\
\hline
24 \\
-24 \\
\hline
0
\end{array}
$$

3. If you get a number other than zero after your last subtraction, this number is your remainder.

Example:
9 divided by 4

$$
\begin{array}{r}
2 \\
4\overline{)9} \\
-8 \\
\hline
1
\end{array}
$$

1 is the remainder.

The answer is 2 r1. This answer can also be written as $2\frac{1}{4}$, since there was one part remaining over out of the four parts needed to make a whole.

▶ WORKING WITH INTEGERS

Remember, an integer is a whole number or its opposite. Here are some rules for working with integers:

Adding
Adding numbers with the same sign results in a sum of the same sign:

(pos) + (pos) = pos   and   • (neg) + (neg) = neg

When adding numbers of different signs, follow this two-step process:

1. Subtract the positive values of the numbers. Positive values are the values of the numbers without any signs.
2. Keep the sign of the number with the larger positive value.

Example:

$-2 + 3 =$

1. Subtract the positive values of the numbers: $3 - 2 = 1$.
2. The number 3 is the larger of the two positive values. Its sign in the original example was positive, so the sign of the answer is positive. The answer is positive 1.

Example:

$8 + -11 =$

1. Subtract the positive values of the numbers: $11 - 8 = 3$.
2. The number 11 is the larger of the two positive values. Its sign in the original example was negative, so the sign of the answer is negative. The answer is negative 3.

Subtracting

When subtracting integers, change all subtraction signs to addition signs and change the sign of the number being subtracted to its opposite. Then follow the rules for addition.

Examples:

$(+10) - (+12) = (+10) + (-12) = -2$

$(-5) - (-7) = (-5) + (+7) = +2$

Multiplying and Dividing

A simple rule for remembering the rules is that if the signs are the same when multiplying or dividing two quantities, the answer will be positive. If the signs are different, the answer will be negative.

(pos) × (pos) = pos	$\frac{(pos)}{(pos)} = $ pos
(pos) × (neg) = neg	$\frac{(pos)}{(neg)} = $ neg
(neg) × (neg) = pos	$\frac{(neg)}{(neg)} = $ pos

Examples:

$(10)(-12) = -120$

$-5 \times -7 = 35$

$\frac{-12}{3} = -4$

$\frac{15}{3} = 5$

▶ SEQUENCE OF MATHEMATICAL OPERATIONS

There is an order in which a sequence of mathematical operations must be performed:

> **P: Parentheses/Grouping Symbols.** Perform all operations within parentheses first. If there is more than one set of parentheses, begin to work with the innermost set and work toward the outside. If more than one operation is present within the parentheses, use the remaining rules of order to determine which operation to perform first.
>
> **E: Exponents.** Evaluate exponents.
>
> **M/D: Multiply/Divide.** Work from left to right in the expression.
>
> **A/S: Add/Subtract.** Work from left to right in the expression.

This order is illustrated by the acronym PEMDAS, which can be remembered by using the first letter of each of the words in the phrase: **P**lease **E**xcuse **M**y **D**ear **A**unt **S**ally.

Example:

$$\frac{(5+3)^2}{4} + 27 = \frac{(8)^2}{4} + 27$$
$$= \frac{64}{4} + 27$$
$$= 16 + 27$$
$$= 43$$

▶ PROPERTIES OF ARITHMETIC

Listed below are several properties of mathematics:

> ► **Commutative Property.** This property states that the order of numbers in a multiplication or addition example can be changed without changing the outcome.
> *Examples:*
> $5 \times 2 = 2 \times 5$
> $5a = a5$
> $b + 3 = 3 + b$
> ► **Associative Property.** This property states that parentheses can be moved to group numbers differently when adding or multiplying without affecting the answer.
> *Examples:*
> $2 + (3 + 4) = (2 + 3) + 4$
> $2(ab) = (2a)b$

▶ **Distributive Property.** When a value is being multiplied by a sum or difference, multiply that value by each quantity within the parentheses and then take the sum or difference.

Examples:

$5(a + b) = 5a + 5b$

$5(100 - 6) = (5 \times 100) - (5 \times 6)$

This second example can be proved by performing the calculations:

$5(94) = 500 - 30$

$470 = 470$

▶ FACTORS AND MULTIPLES

Factors are numbers that can be divided into a larger number without a remainder.

Example:

$12 \div 3 = 4$

The number 3 is, therefore, a factor of the number 12. Other factors of 12 are 1, 2, 4, 6, and 12. The common factors of two numbers are the factors that both numbers have in common.

Example:

The factors of 24 = 1, 2, 3, 4, 6, 8, 12, and 24.

The factors of 18 = 1, 2, 3, 6, 9, and 18.

From the above, you can see that the common factors of 24 and 18 are 1, 2, 3, and 6. From this list it can also be determined that the *greatest* common factor of 24 and 18 is 6. Determining the greatest common factor (GCF) is useful for simplifying fractions.

Example:

Simplify $\frac{16}{20}$.

The factors of 16 are 1, 2, 4, 8, and 16. The factors of 20 are 1, 2, 4, 5, and 20. The common factors of 16 and 20 are 1, 2, and 4. The greatest of these, the GCF, is 4. Therefore, to simplify the fraction, both numerator and denominator should be divided by 4.

$$\frac{16 \div 4}{20 \div 4} = \frac{4}{5}$$

Multiples are numbers that can be obtained by multiplying a number x by a positive integer.

Example:

$5 \times 7 = 35$

The number 35 is, therefore, a multiple of the number 5 and of the number 7. Other multiples of 5 are 5, 10, 15, 20, etc. Other multiples of 7 are 7, 14, 21, 28, etc.

The common multiples of two numbers are the multiples that both numbers share.

Example:
Some multiples of 4 are: 4, 8, 12, 16, 20, 24, 28, 32, 36 . . .
Some multiples of 6 are: 6, 12, 18, 24, 30, 36, 42, 48 . . .

Some common multiples are 12, 24, and 36. From the example above it can also be determined that the *least* common multiple of the numbers 4 and 6 is 12, since this number is the smallest number that appeared in both lists. The least common multiple, or LCM, is used when performing addition and subtraction of fractions to find the least common denominator.

Example: (using denominators 4 and 6 and LCM of 12)

$$\frac{1}{4} + \frac{5}{6} = \frac{1(3)}{4(3)} + \frac{5(2)}{6(2)}$$
$$= \frac{3}{12} + \frac{10}{12}$$
$$= \frac{13}{12}$$
$$= 1\frac{1}{12}$$

► DECIMALS

The most important thing to remember about decimals is that the first place value to the right of the decimal point is the tenths place. The place values are as follows:

1	2	6	8	•	3	4	5	7
THOUSANDS	HUNDREDS	TENS	ONES	DECIMAL POINT	TENTHS	HUNDREDTHS	THOUSANDTHS	TEN THOUSANDTHS

In expanded form, this number can also be expressed as: $1{,}268.3457 = (1 \times 1{,}000) + (2 \times 100) + (6 \times 10) + (8 \times 1) + (3 \times .1) + (4 \times .01) + (5 \times .001) + (7 \times .0001)$.

Adding and Subtracting Decimals

Adding and subtracting decimals is very similar to adding and subtracting whole numbers. The most important thing to remember is to line up the decimal points. Zeros may be filled in as placeholders when all numbers do not have the same number of decimal places.

Examples:
What is the sum of 0.45, 0.8, and 1.36?

$$
\begin{array}{r}
{\scriptstyle 1\ \ 1} \\
0.45 \\
0.80 \\
+\ 1.36 \\
\hline
2.61
\end{array}
$$

Take away 0.35 from 1.06.

$$
\begin{array}{r}
{\scriptstyle 0} \\
\not{1}.{}^{1}06 \\
-\ 0.35 \\
\hline
0.71
\end{array}
$$

Multiplication of Decimals

Multiplication of decimals is exactly the same as multiplication of integers, except one must make note of the total number of decimal places in the factors.

Example:
What is the product of 0.14 and 4.3?
First, multiply as usual:

$$
\begin{array}{r}
4.3 \\
\times\ .14 \\
\hline
172 \\
430 \\
\hline
602
\end{array}
$$

Now, to figure out the answer, 4.3 has one decimal place and .14 has two decimal places. Add in order to determine the total number of decimal places the answer must have to the right of the decimal point. In this problem there are a total of 3 (1 + 2) decimal places. When finished multiplying, start from the right side of the answer, and move to the left the number of decimal places previously calculated.

.602

In this example, 602 turns into .602 since there have to be three decimal places in the answer. If there are not enough digits in the answer, add in zeros in front of the answer until there are enough.

Example:
Multiply 0.03×0.2.

$$\begin{array}{r} .03 \\ \underline{\times .2} \\ 6 \end{array}$$

There are three total decimal places in the problem; therefore, the answer must contain three decimal places. Starting to the right of 6, move left three places. The answer becomes 0.006.

Dividing Decimals

Dividing decimals is a little different from integers for the set-up, and then the regular rules of division apply. It is easier to divide if the divisor does not have any decimals. In order to accomplish that, simply move the decimal place to the right as many places as necessary to make the divisor a whole number. If the decimal point is moved in the divisor, it must also be moved in the dividend in order to keep the answer the same as the original problem. $4 \div 2$ has the same solution as its multiples $8 \div 4$ and $28 \div 14$, etc. Moving a decimal point in a division problem is equivalent to multiplying the numerator and denominator of a fraction by the same quantity, which is the reason the answer will remain the same.

If there are not enough decimal places in the answer to accommodate the required move, simply add zeros until the desired placement is achieved. Add zeros after the decimal point to continue the division until the decimal terminates, or until a repeating pattern is recognized. The decimal point in the quotient belongs directly above the decimal point in the dividend.

Example:
What is $.425\overline{)1.53}$?
First, to make .425 a whole number, move the decimal point three places to the right: 425.
Now move the decimal point three places to the right for 1.53: 1,530.
The problem is now a simple long division problem.

$$\begin{array}{r} 3.6 \\ 425.\overline{)1,530.0} \\ \underline{-1,275}\downarrow \\ 2,550 \\ \underline{-2,550} \\ 0 \end{array}$$

Comparing Decimals

Comparing decimals is actually quite simple. Just line up the decimal points and then fill in zeroes at the end of the numbers until each one has an equal number of digits.

Example:
Compare .5 and .005
Line up decimal points   .5
                                   .005
Add zeroes   .500
                                   .005
Now, ignore the decimal point and consider, which is bigger: 500 or 5?
500 is definitely bigger than 5, so .5 is larger than .005.

▶ FRACTIONS

To do well when working with fractions, it is necessary to understand some basic concepts. Here are some math rules for fractions using variables.

To simplify a fraction:

$$\frac{ac}{bc} = \frac{a}{b}$$

To add or subtract fractions with the same denominator:

$$\frac{a}{b} \pm \frac{c}{b} = \frac{a \pm c}{b}$$

To add or subtract fractions with different denominators:

$$\frac{a}{b} \pm \frac{c}{d} = \frac{ad \pm cb}{bd}$$

To multiply any two fractions:

$$\frac{a}{b} \times \frac{c}{d} = \frac{a \times c}{b \times d}$$

To divide any two fractions:

$$\frac{a}{b} \div \frac{c}{d} = \frac{a}{b} \times \frac{d}{c} = \frac{a \times d}{b \times c}$$

To compare two fractions:

If $\frac{a}{b} = \frac{c}{d}$, then $ad = bc$

If $\frac{a}{b} < \frac{c}{d}$, then $ad < bc$

If $\frac{a}{b} > \frac{c}{d}$, then $ad > bc$

Simplifying Fractions

To simplify fractions, identify the greatest common factor (GCF) of the numerator and denominator and divide both the numerator and denominator by the GCF:

Example:
Simplify $\frac{63}{72}$.
The GCF of 63 and 72 is 9 so divide 63 and 72 each by 9 to simplify the fraction:

$\frac{63 \div 9 = 7}{72 \div 9 = 8}$

$\frac{63}{72} = \frac{7}{8}$

Adding and Subtracting Fractions

To add or subtract fractions with like denominators, just add or subtract the numerators and keep the denominator.

Examples:
$\frac{1}{7} + \frac{5}{7} = \frac{6}{7}$ and $\frac{5}{8} - \frac{2}{8} = \frac{3}{8}$

To add or subtract fractions with unlike denominators, first find the Least Common Denominator, or LCD. The LCD is the smallest number divisible by each of the denominators.

For example, for the denominators 8 and 12, 24 would be the LCD because 24 is the smallest number that is divisible by both 8 and 12: $8 \times 3 = 24$, and $12 \times 2 = 24$.

Using the LCD, convert each fraction to its new form by multiplying both the numerator and denominator by the appropriate factor to get the LCD, and then add or subtract the new numerators.

Example
$\frac{1}{3} + \frac{2}{5} = \frac{1(5)}{3(5)} + \frac{2(3)}{5(3)}$

$= \frac{5}{15} + \frac{6}{15}$

$= \frac{11}{15}$

Multiplication of Fractions

Multiplying fractions is one of the easiest operations to perform. To multiply fractions, simply multiply the numerators and the denominators.

Example:

$$\frac{4}{5} \times \frac{6}{7} = \frac{24}{35}$$

If any numerator and denominator have common factors, these may be simplified before multiplying. Divide the common multiples by a common factor. In the example below, 3 and 6 are both divided by 3 before multiplying.

Example:

$$\frac{\cancel{3}^{1}}{5} \times \frac{1}{\cancel{6}_{2}} = \frac{1}{10}$$

Dividing Fractions

Dividing fractions is equivalent to multiplying the dividend by the **reciprocal** of the divisor. To find the reciprocal of any number, switch its numerator and denominator.

For example, the reciprocals of the following numbers are:

$$\frac{1}{3} \rightarrow \frac{3}{1} = 3$$

$$x \rightarrow \frac{1}{x}$$

$$\frac{4}{5} \rightarrow \frac{5}{4}$$

$$5 \rightarrow \frac{1}{5}$$

When dividing fractions, simply multiply the dividend by the divisor's reciprocal to get the answer.

Example:

(dividend) ÷ (divisor)

$$\frac{1}{4} \div \frac{1}{2}$$

Calculate the reciprocal of the divisor:

$$\frac{1}{2} \rightarrow \frac{2}{1}$$

Multiply the dividend ($\frac{1}{4}$) by the reciprocal of the divisor ($\frac{2}{1}$) and simplify if necessary.

$$\frac{1}{4} \div \frac{1}{2} = \frac{1}{4} \times \frac{2}{1}$$

$$= \frac{2}{4}$$

$$= \frac{1}{2}$$

Comparing Fractions

Sometimes it is necessary to compare the size of fractions. This is very simple when the fractions are familiar or when they have a common denominator.

Examples:

$$\frac{1}{2} < \frac{3}{4} \quad \text{and} \quad \frac{11}{18} > \frac{5}{18}$$

If the fractions are not familiar and/or do not have a common denominator, there is a simple trick to remember. Multiply the numerator of the first fraction by the denominator of the second fraction. Write this answer under the first fraction. Then multiply the numerator of the second fraction by the denominator of the first one. Write this answer under the second fraction. Compare the two numbers. The larger number represents the larger fraction.

Examples:

Which is larger: $\frac{7}{11}$ *or* $\frac{4}{9}$?

$$7 \times 9 = 63 \quad\quad 4 \times 11 = 44$$

$63 > 44$, therefore,

$$\frac{7}{11} > \frac{4}{9}$$

Compare $\frac{6}{18}$ *and* $\frac{2}{6}$.

$$6 \times 6 = 36 \quad\quad 2 \times 18 = 36$$

$36 = 36$, therefore,

$$\frac{6}{18} = \frac{2}{6}$$

▶ PERCENTS

Percents are always "out of 100," so 45% means 45 out of 100. Therefore, to write percents as decimals, move the decimal point two places to the left (to the hundredths place).

Examples:

$$45\% = \frac{45}{100} = 0.45$$

$$3\% = \frac{3}{100} = 0.03$$

$$124\% = \frac{124}{100} = 1.24$$

$$0.9 = \frac{.9}{100} = \frac{9}{1,000} = 0.009$$

To solve percentage problems, determine what information has been given in the problem and fill this information into the following template:

_____ is _____% of _____.

Then translate this information into a one-step equation and solve. In translating, remember that "is" translates to "=" and "of" translates to "×." Use a variable to represent the unknown quantity.

Examples:
A) finding a percentage of a given number
In a new housing development there will be 50 houses and 40% of the houses must be completed in the first stage. How many houses are in the first stage?

1. Translate:
_____ is 40% of 50.
x is .40 × 50.

2. Solve.
$x = .40 × 50$
$x = 20$
Twenty is 40% of 50. There are 20 houses in the first stage.

B) finding a number when a percentage is given
Forty percent of the cars on the lot have been sold. If 24 were sold, how many total cars are there on the lot?

1. Translate:
24 is 40% of _____.
$24 = .40 × x$

2. Solve.
$$\frac{24}{.40} = \frac{.40x}{.40}$$
$60 = x$
24 is 40% of 60. There were 60 total cars on the lot.

C) finding what percentage one number is of another

Matt has 75 employees. He is going to give 15 of them a raise. What percent of the employees will receive a raise?

1. Translate:

15 is _____% of 75.

$15 = x \times 75$

2. Solve.

$\frac{15}{75} = \frac{75x}{75}$

$.20 = x$

$20\% = x$

15 is 20% of 75. 20% of the employees will receive raises.

▶ MEAN, MEDIAN, AND MODE

To find the **average**, or **mean**, of a set of numbers, add all of the numbers together and divide by the quantity of numbers in the set.

Average = (sum of set) ÷ (quantity of set)

Example:

Find the average of 9, 4, 7, 6, and 4.

(9 + 4 + 7 + 6 + 4) ÷ 5 = 30 ÷ 5 = 6

The mean, or average, of the set is 6.

(Divide by 5 because there are 5 numbers in the set.)

To find the **median** of a set of numbers, arrange the numbers in ascending or descending order and find the middle value.

▶ If the set contains an odd number of elements, then simply choose the middle value.
Example:
Find the median of the number set: 1, 5, 4, 7, 2.
First arrange the set in order—1, 2, 4, 5, 7—and then find the middle value. Since there are five values, the middle value is the third one: 4. The median is 4.

▶ If the set contains an even number of elements, simply average the two middle values.
Example:
Find the median of the number set: 1, 6, 3, 7, 2, 8.
First arrange the set in order—1, 2, 3, 6, 7, 8—and then find the middle values, 3 and 6. Find the average of the numbers 3 and 6: $\frac{3+6}{2} = \frac{9}{2} = 4.5$. The median is 4.5.

The **mode** of a set of numbers is the number that appears the greatest number of times.

Example:
For the number set 1, 2, 5, 9, 4, 2, 9, 6, 9, 7, the number 9 is the mode because it appears the most frequently.

Here are some conversions you should be familiar with:

Fraction	Decimal	Percentage
$\frac{1}{2}$	.5	50%
$\frac{1}{4}$	.25	25%
$\frac{1}{3}$	.333 . . .	33.$\overline{3}$%
$\frac{2}{3}$	.666 . . .	66.$\overline{6}$%
$\frac{1}{10}$	.1	10%
$\frac{1}{8}$	.125	12.5%
$\frac{1}{6}$	.1666 . . .	16.$\overline{6}$%
$\frac{1}{5}$	.2	20%

▶ SQUARE ROOTS

The square of a number is the product of the number and itself. For example, in the statement $3^2 = 3 \times 3 = 9$, the number 9 is the square of the number 3. If the process is reversed, the number 3 is the square root of the number 9. The symbol for square root is $\sqrt{}$ and is called a **radical**. The number inside of the radical is called the **radicand**.

Example:
$5^2 = 25$ therefore $\sqrt{25} = 5$

Since 25 is the square of 5, it is also true that 5 is the square root of 25.

Perfect Squares
The square root of a number might not be a whole number. For example, the square root of 7 is 2.645751311. . . . It is not possible to find a whole number that can be multiplied by itself to equal 7. A whole number is a perfect square if its square root is also a whole number.

Examples of perfect squares:

1, 4, 9, 16, 25, 36, 49, 64, 81, 100 . . .

ARITHMETIC PRACTICE

Now it's time to practice these skills. Answer the following 15 questions, and then review the answer explanations that follow.

1. A stockperson just received a shipment of 126 cans of paint that need to be displayed. If each shelf holds 9 cans, how many shelves will he need to display all of the paint?
a. 9 shelves
b. 14 shelves
c. 16 shelves
d. 32 shelves

2. A server at a diner gives 15% of her tips to the busboy. At the end of the night she had made $180 in tips. How much money did she give to the busboy?
a. $1.50
b. $2.70
c. $15.00
d. $27.00

3. Doug works as a cashier at a supermarket. If peaches are on sale for $1.25 per pound, how much will 2.4 pounds cost?
a. $2.00
b. $2.40
c. $3.00
d. $3.25

4. A customer service representative wants to know the average number of phone calls that she receives. At work that day, she records that she received 104 phone calls between the hours of 9:00 A.M. to 5:00 P.M. What is the average number of phone calls she received per hour?
a. 13 calls/hr
b. 15 calls/hr
c. 20 calls/hr
d. 8 calls/hr

5. Juana works at a fruit smoothie stand. Before sales tax, how much will it cost for a large blueberry smoothie with a protein supplement?

SMOOTHIE FLAVORS	SMALL	LARGE	EACH SUPPLEMENT $0.55 EXTRA
Strawberry-Banana	2.65	3.75	Calcium
Peach	2.85	3.95	Protein
Blueberry	2.75	3.85	Vitamin C

    a. $3.30
    b. $4.30
    c. $4.40
    d. $5.50

6. A painter has a service charge of $35 per hour plus the cost of the paint. How much is the total bill for a project if she spent $32 on paint and worked for six hours?
    a. $67
    b. $192
    c. $227
    d. $242

7. A landscaper is hired to plant a row of orange trees along the edge of a client's land which measures 60 yards in length. Each tree needs to be at least 20 feet apart, and each tree must be 20 feet from both edges of the land. What is the maximum number of trees that he can plant?
    a. 2
    b. 4
    c. 8
    d. 10

8. A telemarketer figures that 1 in every 16 calls results in a sale of her company's product. She needs 12 sales to make her bonus. If her next shift is 8 hours straight, how many phone calls per hour will she need to make in order to reach her bonus goal?
    a. 16
    b. 24
    c. 12
    d. 48

9. Mei cleans houses for work and charges $9 per hour. If she worked at a house from 9:30 A.M. until 3:00 P.M., how much should she charge?
    a. $40.50
    b. $49.50
    c. $54.00
    d. $58.50

10. A store is currently having a sale where every frame is 30% off the original price. A customer wants to purchase a frame that was originally priced at $24.90. What is the new sale price?

    a. $7.47

    b. $8.30

    c. $16.60

    d. $17.43

11. A customer buys two sweaters for $37.50 each. Sales tax is 8%. If the customer pays with a hundred dollar bill, how much change should the customer get?

    a. $19.00

    b. $25.00

    c. $81.00

    d. $54.00

12. A carpenter has been hired to install a shelf in a bedroom. The bedroom wall is 12 feet wide and the shelf is 5.6 feet. In order for the shelf to be exactly centered on the wall, how much space should she leave on each side?

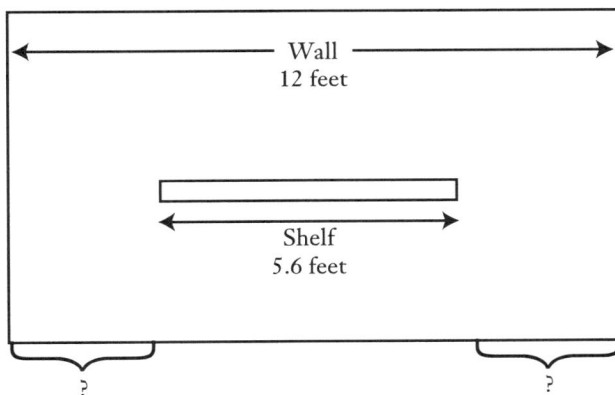

    a. 0.4 ft

    b. 2.3 ft

    c. 3.2 ft

    d. 6.4 ft

13. Victoria is trying to figure out how many cans of paint she will need for her next job. She has three rooms to paint. One room has 480 ft^2 to paint, and the other two rooms have 320 ft^2 each. If each one-gallon can of paint covers 200 ft^2, what is the minimum number of cans that Victoria will have to buy?

    a. 4 cans

    b. 6 cans

    c. 10 cans

    d. 5 cans

14. A cook is trying to decide which vendor he should use to purchase tomatoes. Vendor A charges $25.80 for a 20-pound case of tomatoes. For the same type of tomatoes, Vendor B sells an 8-pound case of tomatoes for $11.60. How much money per pound will the cook save by purchasing from Vendor A?

   a. $0.11

   b. $0.14

   c. $0.16

   d. $0.29

15. A landscaper has six trees to trim. Each tree takes 20 minutes for set-up, two hours for trimming, and 40 minutes for clean-up. He only has two days to complete this project and would like to work the same amount of time each day. How many hours will he need to work each day in order to complete his project on time?

   a. 6 hrs

   b. 8 hrs

   c. 9 hrs

   d. 12 hrs

ANSWERS

1. **b.** Since each shelf can hold 9 cans, divide 126 by 9; $126 \div 9 = 14$

2. **d.** In order to solve the problem, fit the given information into the form: ____ is ____% of ____. In this case, ____ is 15% of $180. Convert 15% to a decimal (0.15) and translate: $x = .15 \times \$180$. Solve by multiplying; $x = \$27$, which is the amount she will give to the busboy.

3. **c.** Multiply the total weight (2.4 lbs) and the sale cost ($1.25) to get the total price; 2.4 lbs × $1.25/lb = $3.00.

4. **a.** The first step is to figure out how many hours the representative worked. 9:00 A.M. to 5:00 P.M. is 8 hours. In order to find the average number of calls per hour, divide the total number of calls (104) by the number of hours worked (8 hours); $104 \div 8 = 13$ calls/hr.

5. **c.** Use the table to answer this problem. A large blueberry smoothie costs $3.85. Since there is a protein supplement, an additional $0.55 is charged. The total is the sum of these two costs; $3.85 + $0.55 = $4.40.

6. **d.** To figure out the painter's total service charge, multiply her hourly rate ($35/hr) and the number of hours she worked (6 hours); $35/hr × 6 hours = $210. Take this total and add the cost of the paint ($32); $210 + $32 = $242.

7. **c.** The length of the plot is 60 yards. Convert this to feet by multiplying by 3. The length of the plot is 180 feet. Break this up into 20-foot sections, since each tree must be 20 feet apart; $180 \div 20 = 9$ sections. Make a picture of the plot of land with the trees drawn in to determine the maximum number of trees. There can be a maximum of 8 trees, since each tree must be 20 feet from the edge.

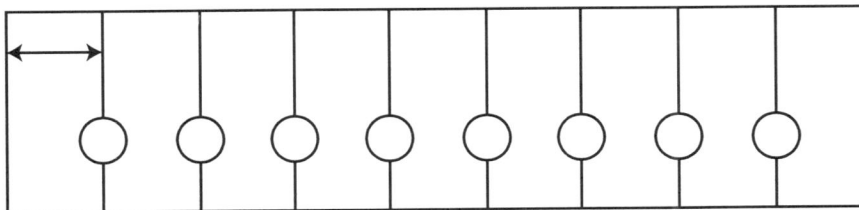

20 ft

60 yds = 180 ft

8. **b.** Since the telemarketer has to make 16 calls in order to achieve 1 sale, multiply the number of sales she needs by 16; 12 sales × 16 calls/sale = 192 calls. She has 8 hours to make 192 calls. Divide 192 by 8 in order to figure out the number of calls per hour; 192 calls ÷ 8 hours = 24 calls/hr. She has to make an average of 24 calls per hour in order to make her bonus goal.

9. **b.** The first step is to figure out how many hours Mei worked; 9:30 A.M. to 3:00 P.M. is 5.5 hours. She works for $9/hr, so 5.5 hours × $9/hr = $49.50.

10. **d.** Translate ____ is 30% of $24.90 into an equation: $x = .30 \times \$24.90$. Solve: $x = \$7.47$, which represents the 30% savings. In order to find the new price of the frame, subtract the 30% savings ($7.47) from the original price ($24.90); $24.90 − $7.47 = $17.43.

11. **a.** Figure out the subtotal cost of the two sweaters; $37.50 × 2 = $75. To calculate the tax, translate ____ is 8% of $75 into an equation: $x = .08 × \$75$. Solve: $x = \$6$ which represents the 8% tax. In order to find the total cost of the sweaters, add the 8% tax ($6) to the original price ($75); $75 + $6 = $81. To figure out the customer's change, subtract the total ($81) from the amount of cash given ($100); $100 − $81 = $19 change.

12. **c.** Subtract the length of the shelf (5.6 ft) from the total length of the wall (12 ft) in order to find out how much extra wall space there is; 12 ft − 5.6 ft = 6.4 ft. Since the shelf has to be centered, the extra 6.4 ft must be distributed evenly on either side; 6.4 ft ÷ 2 = 3.2 ft must be left on each side of the shelf.

13. **b.** The first step is to calculate the total amount of square feet Victoria needs to paint. Two rooms have 320 ft^2 and one has 480 ft^2. The total is the sum of these three rooms; 320 ft^2 + 320 ft^2 + 480 ft^2 = 1,120 ft^2. Next, divide this total by the amount of square feet each gallon covers; 1,120 ft^2 ÷ 200 ft^2/gal = 5.6 gallons. Therefore, the minimum number of cans she will have to purchase is six cans.

14. **c.** In order to compare these two vendors, they both need to be converted to a unit price. The easiest comparison would be price per pound, which is calculated by dividing the total cost by the number of pounds. Vendor A: $25.80 ÷ 20 lbs = $1.29/lb. Vendor B: $11.60 ÷ 8 lbs = 1.45/lb. The question asks how much the cook saves per pound, so subtract the unit price of A from the unit price of B; $1.45 − $1.29 = $0.16.

15. **c.** Add up the various stages in order to figure out the total amount of time he must spend per tree; 20 min + 120 min + 40 min = 180 min (3 hrs). It takes him 3 hours per tree and he has 6 trees to trim; 6 trees × 3 hrs/tree = 18 hrs total. He has 2 days to accomplish 18 hrs of work; 18 hrs ÷ 2 days = 9 hrs/day.

4

Measurement Review

This chapter will review the basics of measurement systems used in the United States and other countries, methods of performing mathematical operations with units of measurement, and the process of converting between different units.

The use of measurement enables a connection to be made between mathematics and the real world. To measure any object, assign a number and a unit of measure. For instance, when a fish is caught, it is often weighed in ounces and its length measured in inches. The following lesson will help you become more familiar with the types, conversions, and units of measurement.

▶ TYPES OF MEASUREMENTS

The types of measurements used most frequently in the United States are listed below:

Units of Length
12 inches (in) = 1 foot (ft)
3 feet = 36 inches = 1 yard (yd)
5,280 feet = 1,760 yards = 1 mile (mi)

Units of Volume
8 ounces* (oz) = 1 cup (c)
2 cups = 16 ounces = 1 pint (pt)
2 pints = 4 cups = 32 ounces = 1 quart (qt)
4 quarts = 8 pints = 16 cups = 128 ounces = 1 gallon (gal)

Units of Weight
16 ounces* (oz) = 1 pound (lb)
2,000 pounds = 1 ton (T)

Units of Time
60 seconds (sec) = 1 minute (min)
60 minute = 1 hour (hr)
24 hours = 1 day
7 days = 1 week
52 weeks = 1 year (yr)
12 months = 1 year
365 days = 1 year

*Notice that ounces are used to measure the dimensions of both volume and weight.

▶ CONVERTING UNITS

When performing mathematical operations, it can be necessary to convert units of measure to simplify a problem. Units of measure are converted by using either multiplication or division:

▶ To convert from a larger unit into a smaller unit, *multiply* the given number of larger units by the number of smaller units in only one of the larger units:
(given number of the larger units) × *(the number of smaller units per larger unit)* = *answer in smaller units.*

For example, to find the number of inches in 5 feet, multiply 5, the number of larger units, by 12, the number of inches in one foot:

5 feet = _____ inches?
5 feet × 12 (the number of inches in a single foot) = 60 inches: $5 \text{ ft} \times \frac{12 \text{ in}}{1 \text{ ft}} = 60 \text{ in}$
Therefore, there are 60 inches in 5 feet.

Example:
Change 3.5 tons to pounds.
3.5 tons = _____ pounds?
$3.5 \text{ tons} \times \frac{2,000 \text{ pounds}}{1 \text{ ton}} = 7,000 \text{ pounds}$
Therefore, there are 7,000 pounds in 3.5 tons.

► To change a smaller unit to a larger unit, *divide* the given number of smaller units by the number of smaller units in only one of the larger units:

$$\frac{\textit{given number of smaller units}}{\textit{the number of smaller units per larger unit}} = \textit{answer in larger units}$$

For example, to find the number of pints in 64 ounces, divide 64, the number of smaller units, by 16, the number of ounces in one pint.

64 ounces = _____ pints?
$$\frac{64 \text{ ounces}}{16 \text{ ounces per pint}} = 4 \text{ pints}$$
Therefore, 64 ounces equals four pints.

Example:
Change 32 ounces to pounds.
32 ounces = _____ pounds?
$$\frac{32 \text{ ounces}}{16 \text{ ounces per pound}} = 2 \text{ pounds}$$
Therefore, 32 ounces equals two pounds.

▷ **BASIC OPERATIONS WITH MEASUREMENT**

You may need to add, subtract, multiply, and divide with measurement. The mathematical rules needed for each of these measurement operations follow.

Addition with Measurements

To add measurements, follow these two steps:

1. Add like units.
2. Simplify the answer by converting smaller units into larger units when possible.

Example:
Add 4 pounds 5 ounces to 20 ounces.

```
    4 lb   5 oz        Be sure to add ounces to ounces.
+        20 oz
    4 lb 25 oz
```

Because 25 ounces is more than 16 ounces (1 pound), simplify by dividing by 16:

```
            1 lb r 9 oz
16 oz)25 oz
      − 16 oz
         9 oz
```

Then add the 1 pound to the 4 pounds:
4 pounds 25 ounces = 4 pounds + 1 pound 9 ounces = 5 pounds 9 ounces

Subtraction with Measurements

To subtract measurements, follow these three steps:
1. Subtract like units if possible.
2. If not, regroup units to allow for subtraction.
3. Write the answer in simplest form.

For example, subtract 6 pounds 2 ounces from 9 pounds 10 ounces.

```
    9 lb 10 oz        Subtract ounces from ounces.
−   6 lb  2 oz        Then subtract pounds from pounds.
    3 lb  8 oz
```

Sometimes, it is necessary to regroup units when subtracting.

Example:

Subtract 3 yards 2 feet from 5 yards 1 foot.

Because 2 feet cannot be taken from 1 foot, regroup 1 yard from the 5 yards and convert the 1 yard to 3 feet. Add 3 feet to 1 foot. Then subtract feet from feet and yards from yards:

$$\begin{array}{r} \overset{4}{\cancel{5}}\ \text{yd}\ \overset{4}{\cancel{1}}\ \text{ft} \\ -\ \underline{3\ \text{yd}\ 2\ \text{ft}} \\ 1\ \text{yd}\ 2\ \text{ft} \end{array}$$

5 yards 1 foot – 3 yards 2 feet = 1 yard 2 feet

Multiplication with Measurements

To multiply measurements, follow these two steps:

1. Multiply like units if units are involved.
2. Simplify the answer.

Example:

Multiply 5 feet 7 inches by 3.

$$\begin{array}{r} 5\ \text{ft}\ 7\ \text{in} \\ \times\ \underline{\qquad\ 3} \\ 15\ \text{ft}\ 21\ \text{in} \end{array}$$

Multiply 7 inches by 3, then multiply 5 feet by 3. Keep the units separate.

Since 12 inches = 1 foot, simplify 21 inches.

15 ft 21 in = 15 ft + 1 ft 9 in = 16 ft 9 in

Example:

Multiply 9 feet by 4 yards.

First, decide on a common unit: either change the 9 feet to yards, or change the 4 yards to feet. Both options are explained below:

Option 1:

To change yards to feet, multiply the number of feet in a yard (3) by the number of yards in this problem (4); 3 feet in a yard × 4 yards = 12 feet.

Then multiply: 9 feet × 12 feet = 108 square feet.

(*Note:* feet × feet = square feet = ft^2)

Option 2:
To change feet to yards, divide the number of feet given (9), by the number of feet in a yard (3); 9 feet ÷ 3 feet in a yard = 3 yards.
Then multiply 3 yards by 4 yards = 12 square yards.
(*Note:* yards × yards = square yards = yd^2)

Division with Measurements

To divide measurements, follow these five steps:

1. Divide into the larger units first.
2. Convert the remainder to the smaller unit.
3. Add the converted remainder to the existing smaller unit, if any.
4. Then divide into smaller units.
5. Write the answer in simplest form.

Example:
Divide 5 quarts 4 ounces by 4.

1. Divide into the larger unit:

$$
\begin{array}{r}
1 \text{ qt r } 1 \text{ qt} \\
4\overline{)5 \text{ qt}} \\
\underline{-\,4 \text{ qt}} \\
1 \text{ qt}
\end{array}
$$

2. Convert the remainder:
1 qt = 32 oz

3. Add remainder to original smaller unit:
32 oz + 4 oz = 36 oz

4. Divide into smaller units:
36 oz ÷ 4 = 9 oz

5. Write answer in simplest form:
1 qt 9 oz

► METRIC MEASUREMENTS

The metric system is an international system of measurement also called the **decimal system**. Converting units in the metric system is much easier than converting units in the English system of measurement. However, making conversions between the two systems is much more difficult. The basic units of the metric system are the meter, gram, and liter. Here is a general idea of how the two systems compare:

METRIC SYSTEM	ENGLISH SYSTEM
1 meter	A meter is a little more than a yard; it is equal to about 39 inches.
1 gram	A gram is a very small unit of weight; there are about 30 grams in one ounce.
1 liter	A liter is a little more than a quart.

Prefixes are attached to the basic metric units listed above to indicate the amount of each unit. For example, the prefix *deci-* means one-tenth ($\frac{1}{10}$); therefore, one decigram is one-tenth of a gram, and one decimeter is one-tenth of a meter. The following six prefixes can be used with every metric unit:

KILO	HECTO	DEKA	DECI	CENTI	MILLI
(k)	(h)	(dk)	(d)	(c)	(m)
1,000	100	10	$\frac{1}{10}$	$\frac{1}{100}$	$\frac{1}{1,000}$

Examples:
  1 hectometer = 1 hm = 100 meters
  1 millimeter = 1 mm = $\frac{1}{1,000}$ meter = .001 meter
  1 dekagram = 1 dkg = 10 grams
  1 centiliter = 1 cL* = $\frac{1}{100}$ liter = .01 liter
  1 kilogram = 1 kg = 1,000 grams
  1 deciliter = 1 dL* = $\frac{1}{10}$ liter = .1 liter
  *Notice that *liter* is abbreviated with a capital letter—"L."

The chart below illustrates some common relationships used in the metric system:

LENGTH	WEIGHT	VOLUME
1 km = 1,000 m	1 kg = 1,000 g	1 kL = 1,000 L
1 m = .001 km	1 g = .001 kg	1 L = .001 kL
1 m = 100 cm	1 g = 100 cg	1 L = 100 cL
1 cm = .01 m	1 cg = .01 g	1 cL = .01 L
1 m = 1,000 mm	1 g = 1,000 mg	1 L = 1,000 mL
1 mm = .001 m	1 mg = .001 g	1 mL = .001 L

▷ CONVERSIONS WITHIN THE METRIC SYSTEM

An easy way to do conversions within the metric system is to move the decimal point either to the right or left because the conversion factor is always ten or a power of ten. Remember, when changing from a large unit to a smaller unit, multiply. When changing from a small unit to a larger unit, divide.

Making Easy Conversions within the Metric System

When multiplying by a power of ten, move the decimal point to the right, since the number becomes larger. When dividing by a power of ten, move the decimal point to the left, since the number becomes smaller.

To change from a large unit to a smaller unit, move the decimal point to the right.

$$\rightarrow$$
$$\text{kilo hecto deka UNIT deci centi milli}$$
$$\leftarrow$$

To change from a small unit to a larger unit, move the decimal point to the left.

SPECIAL TIPS FOR TESTS

An easy way to remember the metric prefixes is to remember the mnemonic: "King Henry Died of Drinking Chocolate Milk." The first letter of each word represents a corresponding metric heading from Kilo down to Milli: "King"—Kilo, "Henry"—Hecto, "Died"—Deka, "of"—original Unit, "Drinking"—Deci, "Chocolate"—Centi, and "Milk"—Milli.

Example:
Change 520 grams to kilograms.

1. Be aware that changing meters to kilometers is going from small units to larger units and, thus, requires that the decimal point move to the left.
2. Beginning at the UNIT (for grams), note that the Kilo heading is three places away. Therefore, the decimal point will move three places to the left.

⌢⌢⌢
k h dk unit d c m

3. Move the decimal point from the end of 520 to the left three places.

520

←

.520

Place the decimal point before the 5.
The answer is 520 grams = .520 kilograms.

Example:
Ron's supply truck can hold a total of 20,000 kilograms. If he exceeds that limit, he must buy stabilizers for the truck that cost $12.80 each. Each stabilizer can hold 100 additional kilograms. If he wants to pack 22,300,000 grams of supplies, how much money will he have to spend on the stabilizers?

1. First, change 22,300,000 grams to kilograms.

⌢ ⌢ ⌢
kg hg dkg g dg cg mg

2. Move the decimal point three places to the left: 22,300,000 g = 22,300.000 kg = 22,300 kg
3. Subtract to find the amount over the limit: 22,300 kg – 20,000 kg = 2,300 kg
4. Because each stabilizer holds 100 kilograms and the supplies exceed the weight limit of the truck by 2,300 kilograms, Ron must purchase 23 stabilizers: 2,300 kg ÷ 100 kg per stabilizer = 23 stabilizers.
5. Each stabilizer costs $12.80, so multiply $12.80 by 23: $12.80 × 23 = $294.40.

MEASUREMENT PRACTICE

Now it's time to practice the skills in this chapter. Answer the following 15 questions, and then review the answer explanations that follow.

1. A carpenter is measuring a piece of plywood. The piece of plywood is 8 feet 7 inches long. He only needs a piece of plywood 7 feet 6 inches long. How many inches does he need to cut off to have the correct length?
   a. 12 inches
   b. 13 inches
   c. 11 inches
   d. 14 inches

2. Edmund is doing repairs on a house and needs to cut a square in the wall for an intercom. Each side of the square needs to be 18 cm. If 1 inch = 2.54 centimeters, how many inches is each side? (Round to the tenths place)
   a. 7.1 in
   b. 9.0 in
   c. 7.8 in
   d. 45.7 in

3. An electrician needs to calculate how many centimeters of electrical tape he needs for a job. If there are 100 centimeters in 1 meter, how many centimeters of electrical tape will he need if he needs 10 meters of tape?
   a. 110 centimeters
   b. 1,000 centimeters
   c. 10,000 centimeters
   d. 1,100 centimeters

4. Teresa is following the directions for mixing cement. According to the directions, she needs 13 gallons of water for the amount of cement mix she is using. She only has a container that measures liters. If 1 gallon = 3.8 liters, how many liters will she need?
   a. 3.4 L
   b. 494 L
   c. 39.0 L
   d. 49.4 L

5. A customer service representative is asked how many miles the company is from Town A. She knows that the company is between Town A and Town B. She also knows that it is 16.6 miles between Town A and Town B, and that Town B is 7.2 miles from the company. What is the correct answer to the customer's question?

    a. 7.2 miles

    b. 8.3 miles

    c. 9.4 miles

    d. 23.8 miles

6. Mishka took a water measurement that read 174 milliliters. All of his records need to be in liters. Based on the 174 milliliters reading, how many liters should he record?

    a. 174 L

    b. 0.174 L

    c. 1.74 L

    d. 0.0174 L

7. Tess works at a fabric store and is helping a customer who wants to buy two yards of French ribbon. If the price of the ribbon is $0.75 per foot, what is the total cost to the customer?

    a. $1.50

    b. $3.00

    c. $4.30

    d. $4.50

8. Lee is constructing a frame for a poster. The poster is 24 inches wide by 32 inches long and he wants the frame to expand beyond the poster 2 inches on every side. What will the perimeter of the frame be?

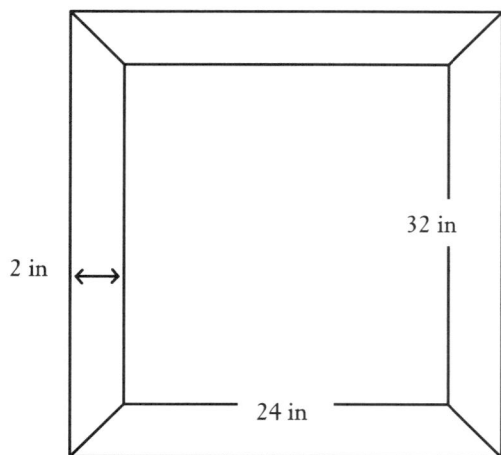

    a. 112 in

    b. 116 in

    c. 120 in

    d. 128 in

9. Horatio is asked by his supervisor to consolidate three bins of apples into one. The first bin has 5 lbs 6 oz of apples, the second has 7 lbs 12 oz, and the third has 14 lbs 4 oz. After he puts all the apples into one bin, what is the total weight of the apples? (1 pound = 16 ounces)

a. 26 lbs 6 oz

b. 27 lbs 6 oz

c. 28 lbs 2 oz

d. 27 lbs 12 oz

10. A carpenter needs to center a headboard in a room against the wall. How much room should she leave on either side of the headboard if the wall is 12 feet 2 inches wide and the headboard is 6 feet 6 inches wide?

a. 5'6"

b. 5'8"

c. 2'8"

d. 2'10"

11. A server is setting up tables for a private party in the back room of the restaurant where he works. He has two 3 ft × 9 ft tables that he must put together to make 1 large table. How much greater will the perimeter be if he chooses set-up A instead of set-up B?

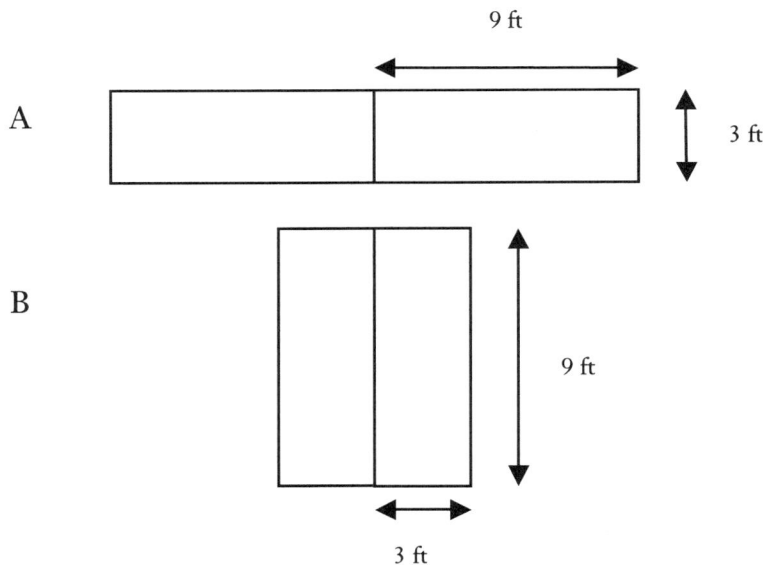

9 ft

A

3 ft

B

9 ft

3 ft

a. 3 ft

b. 6 ft

c. 9 ft

d. 12 ft

12. Ricardo is hired to build a wooden fence around the lot below. What is the perimeter of this lot?

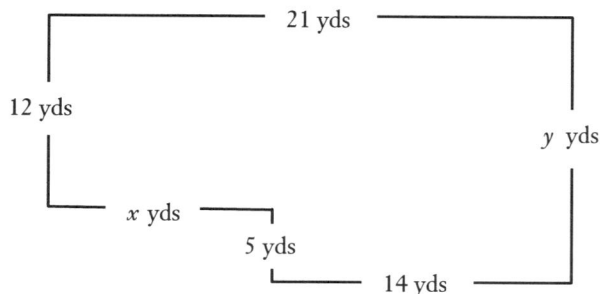

    a. 54 yds
    b. 66 yds
    c. 76 yds
    d. 252 yds

13. A mason worker has been hired to tile a rectangular area of a patio that measures 2 yds × 4 yds. He has been instructed to tile the entire area with no space left in between the tiles. Each tile is square and measures 3 in × 3 in. How many tiles will he need in order to cover the area of the patio?
    a. 96 tiles
    b. 864 tiles
    c. 1,152 tiles
    d. 3,456 tiles

14. A telemarketer is arranging the delivery of a large piece of exercise equipment to a customer. The moving company charges $0.40 per pound and $0.35 per mile. The exercise machine weighs 215 lbs and the customer lives 88 kilometers away. How much will the cost of delivery be? (1 mi = 1.6 km)
    a. $105.25
    b. $116.80
    c. $135.28
    d. $174.00

15. Erika is trying to decide what size shades she should get to cover a window in a client's house that she is decorating. The window's length is twice as long as the window's width. If the width is 0.90 m, what dimensions, in inches, should her shade be (1 inch = 2.54 cm)? Round your answer to the nearest tenth.

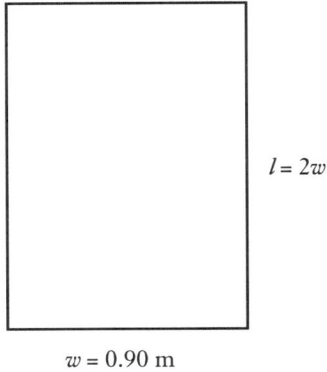

$l = 2w$

$w = 0.90$ m

a.   35.4 in × 17.7 in
b.   35.4 in × 70.9 in
c.   90 in × 180 in
d.   228.6 in × 457.2 in

ANSWERS

1. **b.** This is a subtraction problem. Subtract 7 feet 6 inches from 8 feet 7 inches; 7 inches – 6 inches = 1 inch, and 8 feet – 7 feet = 1 foot. Therefore, 1 foot 1 inch needs to be cut. However, the question asks for the answer in inches; 1 foot = 12 inches, so add 12 inches and 1 inch, and the answer is 13 inches.

2. **a.** To calculate inches, divide 18 cm by 2.54 cm/in; 18 cm ÷ 2.54 ≈ 7.0866 inches. Rounding to the tenths place, 7.0866 ≈ 7.1 inches.

3. **b.** If there are 100 centimeters in 1 meter, then to find how many centimeters are in 10 meters, multiply 100 by 10; 100 cm/m × 10 m = 1,000 centimeters.

4. **d.** There are 3.8 L per 1 gallon. To figure out how many liters Teresa will need for her mix, convert gallons to liters by multiplying 13 gal × 3.8 L/gal = 49.4 L.

5. **c.** To calculate the distance between Town A and the company, take the total distance (16.6 miles) and subtract the distance from the company to Town B; 16.6 miles – 7.2 miles = 9.4 miles.

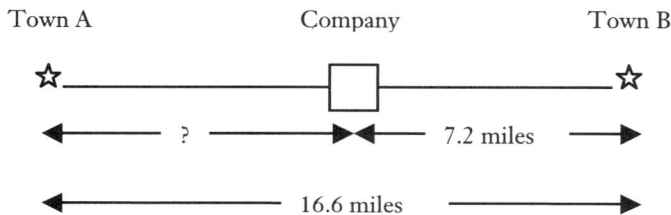

6. **b.** There are 1,000 milliliters in 1 liter. To calculate how many liters 174 milliliters are, divide 174 by 1,000; 174 mL ÷ 1,000 mL/L= 0.174 L.

7. **d.** First convert yards to feet. Since there are 3 ft in a yard, the customer wants to buy 6 ft of ribbon: 2 yds × 3 ft/yd = 6 ft. Multiply 6 ft by the cost of ribbon per foot to get the total price; 6 ft × $0.75/ft = $4.50.

8. **d.** To calculate the perimeter, add up all four sides of the rectangle. The length of the frame will be the length of the poster (32 in) plus two more inches on *each* side; 32 in + 2 in + 2 in = 36 in. The width of the frame is the width of the poster (24 in) plus two inches on *each* side; 24 in + 2 in + 2 in = 28 in. To find the perimeter add the two widths and two lengths together; 28 in + 28 in + 36 in + 36 in = 128 in.

9. **b.** To find the total weight, add up the pounds and ounces separately. The sum of the pounds is 5 lbs + 7 lbs + 14 lbs = 26 lbs. Next, add the ounces from each bin, 6 oz + 12 oz + 4 oz = 22 oz. Since there are 16 oz in 1 lb, 22 oz is equal to 1 lb and 6 oz. Add this number to the total pounds from the first step; 26 lbs + 1 lb 6oz = 27 lbs 6 oz.

10. **d.** First, convert the dimensions into inches; wall = 12 ft 2 in. There are 12 inches in a foot, so 12ft × 12 in/ft = 144 inches. Don't forget to add the extra two inches from the original measurement; wall = 144 in + 2 in = 146 in; headboard = 6ft 6 in; 6ft × 12in/ft = 72 in; headboard = 72 in + 6 in = 78 in. Now subtract the width of the headboard from the width of the width of the wall; 146 in – 78 in = 68 in. There needs to be an equal amount of room on either side of

the headboard, so divide this number by 2; 68 in ÷ 2 = 34 in. Convert this value back into ft; 34in ÷ 12in/ft = 2 ft with a remainder of 10 inches.

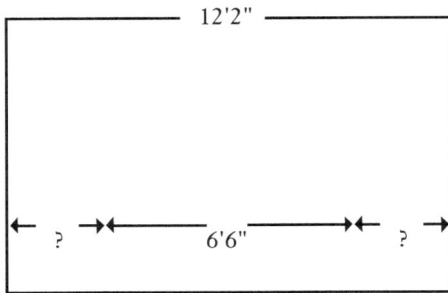

11. d. Based on the given values for the length and width of the tables, the measurements of each side can be calculated. To find the perimeter of each set-up, add the lengths of each side making sure not to include the value of the sides that do not contribute to the outside of the set-up. Add up the four sides to get the perimeter; A = 18 + 3 + 18 + 3 = 42 ft; B = 6 + 9 + 6 + 9 = 30 ft. Find the difference by subtracting; 42 ft − 30 ft = 12 ft.

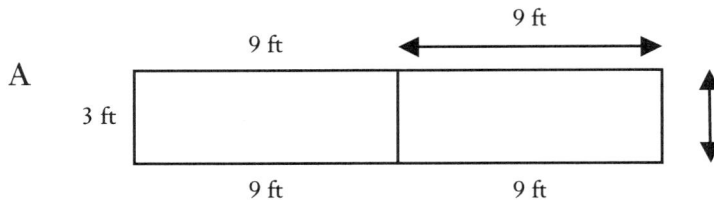

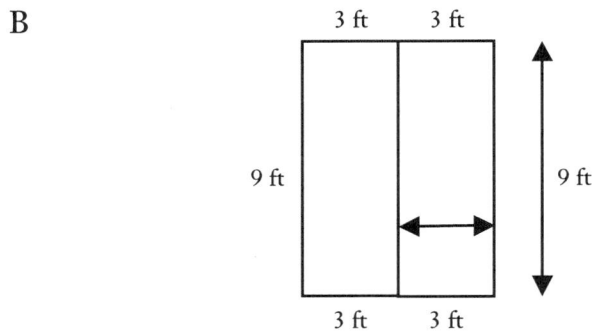

12. c. To find the perimeter, add the length of each side. There are six sides total, but only four values are given. The missing measurements can be calculated. The value for y is the sum of the two length measurements; 12 yds + 5 yds = 17 yds. The value for x is found by subtracting the width of the value given from the entire width; 21 yds – 14 yds = 7 yds. To find the perimeter, add up the values of all 6 sides; 21 + 17 + 14 + 5 + 7 + 12 = 76 yds.

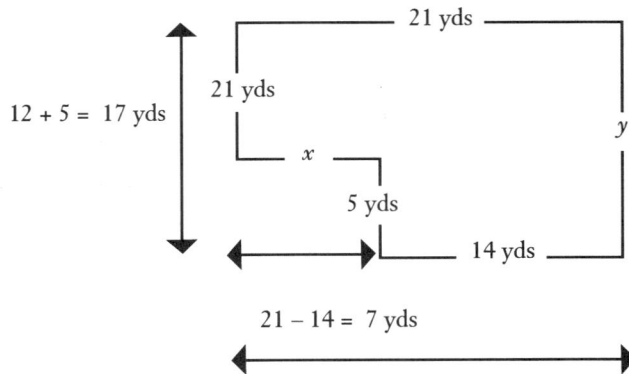

13. c. Convert the values to a single unit. There are 3 ft in a yd, so 2 yds × 3 ft/yd = 6 feet, and 4 yds × 3 ft/yd = 12 feet. Now, convert feet to inches. There are 12 inches per 1 foot, so 6 feet × 12 in/ft = 72 in, and 12 feet × 12 in/ft = 144 in. Therefore, the patio is 72 inches by 144 inches. To figure out the area of the patio, multiply the length times the width; 72 in × 144 in = 10,368 in^2. The surface area of each tile is 3″ × 3″ = 9 in^2. Divide the total surface area of the patio by the surface area of an individual tile to figure out how many tiles are needed. Thus, 10,368 in^2 ÷ 9 in^2/tile = 1,152 tiles.

14. a. First calculate the weight cost; 215 lbs × \$0.40/lb = \$86. The next step is to calculate the distance cost. Convert the 88 km into miles; 88 km ÷ 1.6km/mile = 55 miles. The company charges \$0.35 per mile; 55 miles × \$0.35 = \$19.25. Add the weight cost to the distance cost to get the total; \$86 + \$19.25 = \$105.25.

15. b. First, convert 0.90 m to cm; .90 m × 100 cm/m = 90 cm. So, the width is 90 cm and the length is twice the width, or 180 cm (90 cm × 2). Since the answer must be in inches, the measurements must be converted from centimeters to inches. There are 2.54 cm for every 1 inch; width = 90 cm ÷ 2.54 cm/in = 35.4 inches, to the nearest tenth.; length = 180 cm ÷ 2.54 cm/in = 70.9 inches, to the nearest tenth.

Algebra Review

This chapter will help in mastering algebraic equations by reviewing variables, cross multiplication, algebraic fractions, reciprocal rules, and exponents. Although algebra covers several other topics, these will be most helpful on the job.

Algebra is arithmetic using letters, called **variables**, in place of numbers. By using variables, the general relationships among numbers can be easier to see and understand.

▶ EQUATIONS

An equation is solved by finding the value of an unknown variable. A variable is a letter that represents an unknown number. Variables are frequently used in equations, formulas, and in mathematical rules to help illustrate numerical relationships.

When a number is placed next to a variable, indicating multiplication, the number is said to be the **coefficient** of the variable.

Example:

8c 8 is the coefficient to the variable *c*

6ab 6 is the coefficient to both variables, *a* and *b*

If two or more terms have exactly the same variable(s), and these variables are raised to exactly the same exponents, they are said to be **like terms**. Like terms can be added and subtracted.

Examples:

$7x + 3x = 10x$

$6y^2 - 4y^2 = 2y^2$

$3cd^2 + 5c^2d \rightarrow$ cannot be added. Since the exponent of 2 is on *d* in $3cd^2$ and on *c* in $5c^2d$, they are not like terms.

The process of adding and subtracting like terms is called **combining** like terms. It is important to combine like terms carefully, making sure that **the variables are exactly the same**.

▶ SIMPLE RULES FOR WORKING WITH EQUATIONS

1. The equal sign separates an equation into two sides.
2. Whenever an operation is performed on one side of the equation, the same operation must be performed on the other side of the equation.
3. The first goal is to get all of the variables on one side of the equation and all of the numbers on the other side of the equation. This is accomplished by "undoing" the operations that are attaching numbers to the variable, thereby isolating the variable.
4. The final step will often be to divide each side by the coefficient, the number in front of the variable, leaving the variable alone and equal to a number.

Example:

$$5m + 8 = 48$$
$$-8 = -8$$
$$\frac{5m}{5} = \frac{40}{5}$$
$$m = 8$$

"Undo" the addition of 8 by subtracting 8 from both sides of the equation. Then "undo" the multiplication by 5 by dividing by 5 on both sides of the equation. The variable, *m*, is now isolated on the left side of the equation, and its value is 8.

▶ CHECKING EQUATIONS

To check an equation, substitute the value of the variable into the original equation.

Example:
To check the previous equation, substitute the number 8 for the variable *m* in $5m + 8 = 48$.
$5(8) + 8 = 48$
$40 + 8 = 48$
$48 = 48$
Because this statement is true, the answer $m = 8$ must be correct.

▶ ASSIGNING VARIABLES IN WORD PROBLEMS

It may be necessary to create and assign variables in a word problem. To do this, first identify an "unknown" and a "known." The value of the "known" may not be numerical, but the problem should reveal something about its value.

Examples:
1. Mary Fran has worked three more hours than Jen.
Unknown: Jen's hours = x
Known: Mary Fran has three more hours = $x + 3$
Therefore,
Jen's hours = x and Mary Fran's hours = $x + 3$

2. There are twice as many paperback books as hardcover books.
Unknown: number of hardcover books = x
Known: number of paperback books is twice the number of hardcover books = $2x$
3. Todd has assembled five more than three times the number of cabinets that Andrew has assembled.
Unknown: the number of cabinets Andrew has assembled = x
Known: the number of cabinets Todd has assembled is five more than three times the number Andrew has assembled = $3x + 5$

▶ CROSS MULTIPLYING

Since algebra uses percents and proportions, it is necessary to learn how to cross multiply. You can solve an equation that sets one fraction equal to another by **cross multiplication**. Cross multiplication involves setting the cross products of opposite pairs of terms equal to each other.

SPECIAL TIPS FOR TESTS

If you have to take a math test for your job, remember these tips.

- If time permits, be sure to check all equations.
- Be careful to answer the question that is being asked. Sometimes, this involves solving for a variable and then performing an additional operation. Example: If the question asks the value of $x - 2$, and you find that $x = 2$, the correct answer to the question is not 2, but $2 - 2$. Thus, the answer is 0.

Example:

$$\frac{x}{10} = \frac{70}{100}$$
$$100x = 700$$
$$\frac{100x}{100} = \frac{700}{100}$$
$$x = 7$$

▶ ALGEBRAIC FRACTIONS

Algebraic fractions are very similar to fractions in arithmetic.

Example:
A hotel currently has only one-fifth of their rooms available. If x represents the total number of rooms in the hotel, find an expression for the number of rooms that will be available if another tenth of the total rooms are reserved.

Solution
Since x represents the total number of rooms, $\frac{x}{5}$ represents the number of available rooms. One tenth of the total rooms in the hotel would be represented by the fraction $\frac{x}{10}$. To find the new number of available rooms, find the difference: $\frac{x}{5} - \frac{x}{10}$.
Write $\frac{x}{5} - \frac{x}{10}$ as a single fraction.
Just like in arithmetic, the first step is to find the LCD of 5 and 10, which is 10. Then change each fraction into an equivalent fraction that has 10 as a denominator.

$$\frac{x}{5} - \frac{x}{10} = \frac{x(2)}{5(2)} - \frac{x}{10}$$
$$= \frac{2x}{10} - \frac{x}{10}$$
$$= \frac{x}{10}$$

So, $\frac{x}{10}$ rooms will be available after another tenth of the rooms are reserved.

▶ RECIPROCAL RULES

There are special rules for the sum and difference of reciprocals. The reciprocal of 3 is $\frac{1}{3}$ and the reciprocal of x is $\frac{1}{x}$.

▶ If x and y are not 0, then $\frac{1}{x} + \frac{1}{y} = \frac{y}{xy} + \frac{x}{xy} = \frac{y+x}{xy}$.

▶ If x and y are not 0, then $\frac{1}{x} - \frac{1}{y} = \frac{y}{xy} - \frac{x}{xy} = \frac{y-x}{xy}$.

▶ EXPONENTS

An exponent indicates the number of times a base is used as a factor to attain a product.

Example:
In the expression 2^5, 2 is the base and 5 is the exponent. Therefore, 2 should be used as a factor 5 times to attain a product.
$2^5 = 2 \times 2 \times 2 \times 2 \times 2 = 32$
It is possible for a variable to have an exponent: b^n.
The "b" represents a number that will be a factor to itself "n" times.

Example:
Find the value of b^n if $b = 5$ and $n = 3$.
$b^n = 5^3 = 5 \times 5 \times 5 = 125$
Don't let the variables become intimidating. Most expressions are very simple once numbers have been substituted for the variables. And remember, any base to the zero power is always 1.

Examples:
$5^0 = 1$ $70^0 = 1$ $29,874^0 = 1$

ALGEBRA PRACTICE

Now it's time to practice your skills. Answer the following 15 questions, and then review the answer explanations that follow.

1. To calculate her fee, an electrician uses the expression $30n$ where n is the number of hours worked to calculate her fee. What is her fee if she works 3 hours?
   a. $33
   b. $90
   c. $60
   d. $27

2. $F = 25A$ is the formula to find the number of pounds of fertilizer needed to cover a lawn. A is the area of the lawn in square feet. How many pounds of fertilizer (F) are needed to cover a lawn with an area (A) of 500 square feet?
   a. 12,500 pounds
   b. 2,000 pounds
   c. 525 pounds
   d. 100 pounds

3. To calculate profit (P), a retailer uses the formula $P = I - E$. If the income (I) is $3,500 and the expenses ($E$) are $1,700, what is the profit?
   a. $5,200
   b. $1,800
   c. $800
   d. $2,200

4. A carpenter is using the diagram below to build a porch. The width of the porch, x, is 12 feet. What is the length?

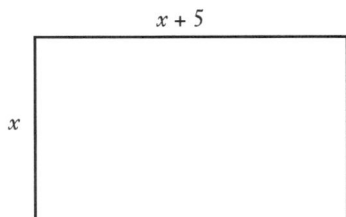

   a. 17 feet
   b. 7 feet
   c. 10 feet
   d. 60 feet

5. A retailer is having a storewide sale. Every item is 10% off its original price. If p = the original price, which expression below represents the sale price of each item?
   a. $p - 0.10p$
   b. $p + 0.10p$
   c. $0.10p$
   d. $p + 0.10$

6. The manager of a cafeteria uses algebraic expressions to calculate the number of servings she can prepare with the amount of food she has in stock. If each can of spaghetti sauce contains 20 servings, which expression below represents the number of servings in n cans of spaghetti sauce?
   a. $20 + n$
   b. $20 - n$
   c. $20 \times n$
   d. $20 \div n$

7. A post is needed for every four feet of fencing. Which expression can be used to calculate the number of posts needed to fence a rectangular area if f is the number of feet of fencing?
   a. $f + 4$
   b. $f - 4$
   c. $f \times 4$
   d. $f \div 4$

8. Joon is in charge of ordering bulk food for a sandwich shop. He purchases pickles on a weekly basis but recently switched vendors and needs to recalculate his order for the upcoming week. He estimates that they use 8 lbs of pickles each day (they are open 6 days a week). The new vendor only sells pickles in 20 lb containers. According to his estimate, what is the minimum number of containers Joon must order in order to guarantee he has enough pickles for the week?
   a. 3 containers
   b. 4 containers
   c. 5 containers
   d. 6 containers

9. A restaurant charges 15% of the total bill for delivery. Which expression represents the delivery charge on a total bill, t?
   a. $15 + t$
   b. $0.15 + t$
   c. $0.15t$
   d. $15t$

10. To calculate his weekly earnings (E), a salesperson uses the formula $E = 0.22s + 150$, where s is his total sales. What did he earn last week if his sales were $2,200?

   a. $590

   b. $634

   c. $980

   d. $2,350

11. A shipping company charges $2.95 for the first 2 pounds and then $0.30 for every ounce over 2 pounds. The cost for shipping a package is found using the formula $C = 2.95 + 0.30w$, where w is the number of ounces over 2 pounds. What is the cost to ship a package that weighs 2 pounds 14 ounces?

   a. $3.09

   b. $2.14

   c. $7.15

   d. $4.55

12. Rick works at a supermarket and has been instructed to set up a sale display with canned kidney beans. Each can weighs $\frac{4}{5}$ lb. A sign on the shelf warns that each shelf cannot support more than 30 lbs of weight. What is the maximum number of cans that Rick can place on a shelf without exceeding the weight limit?

   a. 37 cans

   b. 38 cans

   c. 30 cans

   d. 25 cans

13. A cook uses the formula $A = \frac{3}{2}n$, where n is the number of customers, to determine how many cups of mashed potatoes are needed. How many cups of mashed potatoes does he need for 90 customers?

   a. 135 cups

   b. 270 cups

   c. 90 cups

   d. 145 cups

14. Long distance phone calls cost ten cents for the first minute plus four cents for each additional minute. Which expression below represents the cost of a phone call in cents? Let m equal the total number of minutes.

   a. $m + 40$

   b. $10 + 4(m - 1)$

   c. $10m + 4m$

   d. $(10 + m)(4 + m)$

15. Ohm's Law states $I = \frac{E}{R}$ where I is the current in amperes, E is the electromotive force in volts, and R is the resistance in ohms. Solve Ohm's Law for E.

    a. $E = IR$

    b. $E = \frac{I}{R}$

    c. $E = I + R$

    d. $E = ERI$

ANSWERS

1. **b.** Replace n with 3. Complete the multiplication: $30 \times 3 = 90$. Her fee is $90.

2. **a.** Multiply 25 by the area of the lawn, 500 square feet; $25 \times 500 = 12,500$; 12,500 pounds are needed.

3. **b.** Replace I with $3,500 and E with $1,700 and subtract; $3,500 - $1,700 = $1,800.

4. **a.** The length is $x + 5$. Replace x with 12; $12 + 5 = 17$ feet.

5. **a.** Subtract 10% of the original price, p, from the original price. To find 10% of the original price, multiply the decimal equivalent of 10% (0.10) by the original price, p: $0.10p$. Subtract this amount from the original price ($p - 0.10p$).

6. **c.** The number of cans, n, must be multiplied by the number of servings in each can, 20. The algebraic expression which represents this calculation is $20 \times n$.

7. **d.** Since a post is needed for every four feet of fencing, divide the number of feet of fencing, f, by four to find the number of posts needed. The expression is $f \div 4$.

8. **a.** The sandwich shop uses 8 lbs of pickles per day. They are open 6 days a week, so to find the amount of pickles they use in a week, multiply 8 lbs/day $\times$ 6 days/wk = 48 lbs/wk. Since the pickles are sold in 20 lb containers, Joon needs to buy 3 containers to cover them for the week, even though there may be 12 lbs left over.

9. **c.** Multiply the total bill, t, by the decimal equivalent of 15%, 0.15. This expression is $0.15t$.

10. **b.** Replace s with $2,200. The equation becomes $E = 0.22 \times 2,200 + 150$. The rules for order of operations state that multiplication is always done before addition. Therefore, complete the multiplication ($0.22 \times 2,200 = 484$) first, then add 150 to the result; $484 + 150 = $634.

11. **c.** Replace w with 14 since the package weighs 14 ounces over 2 pounds. The formula is then $C = 2.95 + 0.30 \times 14$. The order of operations states that multiplication is always done before addition. Multiply 0.30 by 14 ($0.30 \times 14 = 4.20$) then add the result to 2.95; $2.95 + 4.20 = $7.15.

12. **a.** First, start by converting $\frac{4}{5}$ to a decimal; 4 divided by 5 = 0.8. Since each can weighs 0.8 lbs and the shelf can hold 30 lbs, divide 30 by 0.8 to calculate how many cans can be placed; 30 lbs $\div$ 0.8 lbs/can = 37.5 cans. Since you can't put 0.5 of a can on the shelf, the maximum number of cans that can be set on the shelf without exceeding the 30 pound weight limit is 37 cans.

13. **a.** Replace n with 90 in the formula: $A = \frac{3}{2} \times 90$. Calculate the right-hand side of the equation recalling that $90 = \frac{90}{1}$. Multiply the fractions by multiplying the numerators (tops) and then multiplying the denominators (bottoms); $\frac{3}{2} \times \frac{90}{1} = \frac{270}{2} = 135$; 135 cups of mashed potatoes are needed.

14. **b.** The cost of a phone call is ten cents plus four cents multiplied by one less than the total number of minutes (the first minute costs ten cents, so it is not multiplied by four cents). This translates into $10 + 4(m - 1)$; $m - 1$ is in parentheses because it must be calculated before multiplying by four cents.

15. **a.** To solve $I = \frac{E}{R}$ for E, multiply both sides of the equation by R. The Rs cancel each other out on the right side of the equation, so the equation becomes $IR = E$. The equation can be flipped around to become $E = IR$, which is choice **a.**

6

Geometry Review

This chapter will familiarize you with the properties of angles, lines, polygons, triangles, and circles, as well as the formulas for area, volume, and perimeter.

Geometry is the study of shapes and the relationships among them. The geometry that is necessary to function in the workforce is fundamental and practical. Basic concepts in geometry will be detailed and applied in this section. The study of geometry always begins with a look at basic vocabulary and concepts. Therefore, following is a list of definitions and important formulas.

▶ GEOMETRY TERMS

Area	the space inside a two-dimensional figure
Circumference	the distance around a circle
Chord	a line segment that goes through a circle, with its endpoints on the circle

Diameter	a chord that goes directly through the center of a circle—the longest line that can be drawn in a circle
Hypotenuse	the longest leg of a right triangle, always opposite the right angle
Leg	either of the two sides of a right triangle that make the right angle
Perimeter	the distance around a figure
Radius	a line from the center of a circle to a point on the circle (half of the diameter)
Surface Area	the sum of the areas of all of a three-dimensional figure's faces
Volume	the space inside a three-dimensional figure

▶ GEOMETRY FORMULAS

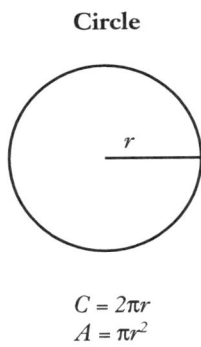

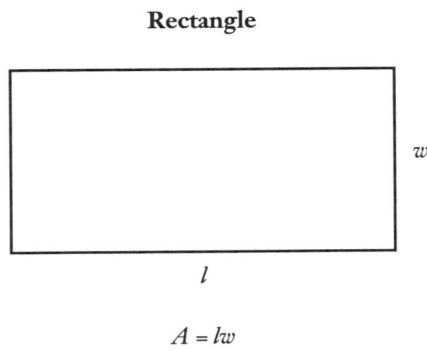

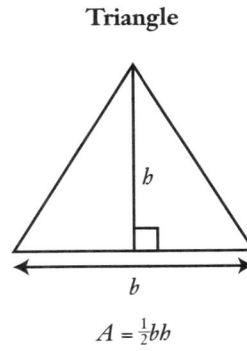

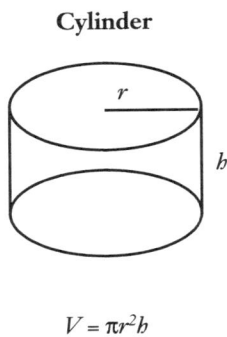

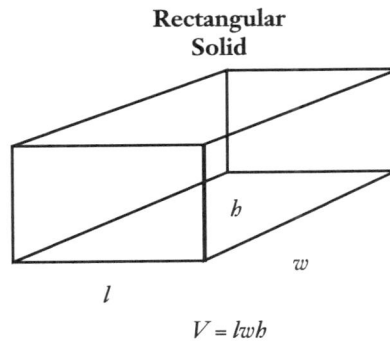

Circle

$C = 2\pi r$
$A = \pi r^2$

Rectangle

$A = lw$

Triangle

$A = \frac{1}{2}bh$

Cylinder

$V = \pi r^2 h$

Rectangular Solid

$V = lwh$

C	=	Circumference	w =	Width
A	=	Area	h =	Height
r	=	Radius	v =	Volume
l	=	Length	b =	Base

Perimeter	the sum of all the sides of a figure
Circumference	$2\pi r$, or πd
Area of a rectangle	$A = bh$

Area of a triangle $A = \frac{1}{2}bh$

It is important to note that the height of a triangle is not necessarily one of the sides of the triangle. The correct height for the following triangle is 8, not 10. The height will always be associated with a line (called an **altitude**) that comes from one **vertex** (angle) to the opposite side and forms a **right angle** (signified by the box). In other words, the height must always be **perpendicular** to (form a right angle with) the base.

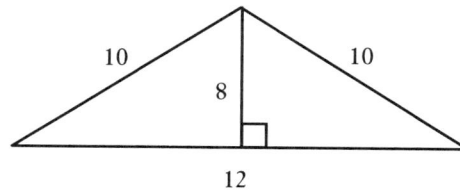

Area of a circle	$A = \pi r^2$
Volume of a cylindrical solid	$A = \pi r^2 h$
Volume of a rectangular solid	$V = lwh$

Key:

π = a ratio used for circles. Most of the time it is okay to approximate π with 3.14. Most calculators have a π key.

r = radius of a circle

d = diameter of a circle

b = base of a figure, primarily for triangles and rectangles

h = height of a figure

l = length of a figure

w = width of a figure, primarily for rectangular objects

BASIC GEOMETRIC FACTS

▶ The sum of the angles in a triangle is 180 degrees.
▶ A circle has a total of 360 degrees.

SHAPE	NUMBER OF SIDES
Circle	0
Triangle	3
Quadrilateral (square/rectangle)	4
Pentagon	5
Hexagon	6
Heptagon	7
Octagon	8
Nonagon	9
Decagon	10

▶ ANGLES

An angle is formed by an endpoint, or vertex, and two rays. A ray includes an endpoint and a line shooting off in one direction.

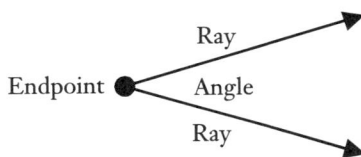

Classifying Angles

Angles can be classified into the following categories: acute, right, obtuse, and straight.

▶ An **acute angle** is an angle that measures between 0 and 90 degrees.

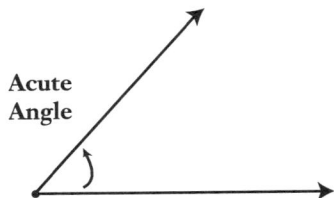

▶ A **right angle** is an angle that measures exactly 90 degrees. A right angle is symbolized by a square at the vertex.

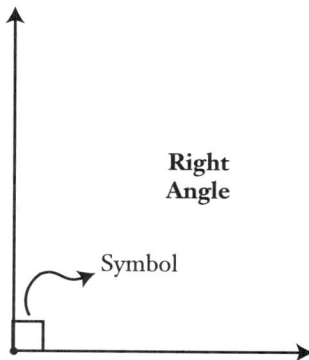

Right Angle

Symbol

▶ An **obtuse angle** is an angle that measures more than 90 degrees, but less than 180 degrees.

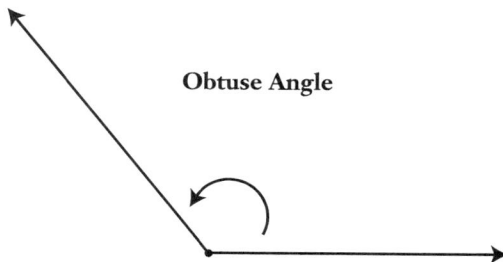

Obtuse Angle

▶ A **straight angle** is an angle that measures 180 degrees. Thus, both of its sides form a line.

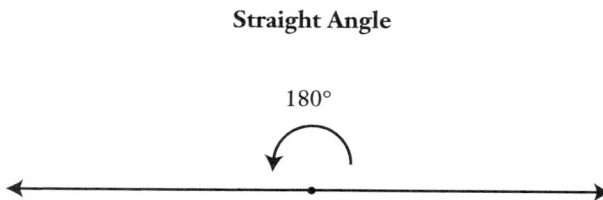

Straight Angle

180°

Complementary Angles

Two angles are **complementary** if the sum of their measures is equal to 90 degrees.

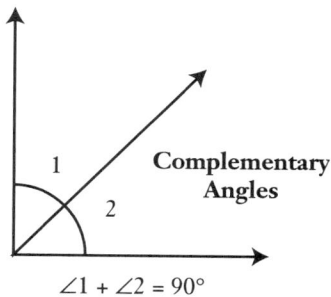

1

2

Complementary Angles

$\angle 1 + \angle 2 = 90°$

Supplementary Angles

Two angles are **supplementary** if the sum of their measures is equal to 180 degrees.

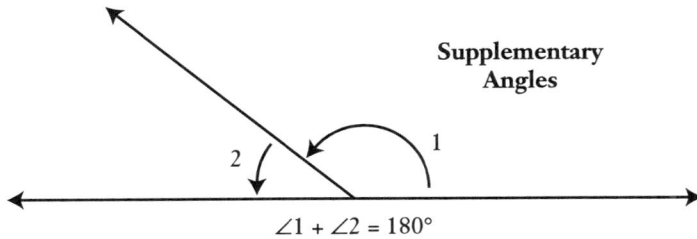

Supplementary Angles

$\angle 1 + \angle 2 = 180°$

► TRIANGLES

Angles of a Triangle

The sum of the measures of the three angles in a triangle always equals 180 degrees.

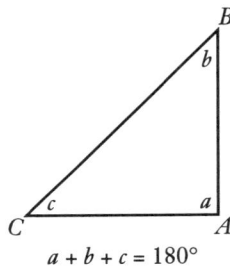

$a + b + c = 180°$

Exterior Angles

An exterior angle can be formed by extending a side from any of the three vertices of a triangle. Here are some rules for working with exterior angles:

► An exterior angle and an interior angle that share the same vertex are supplementary. In other words, exterior angles and interior angles form straight lines with each other.
► An exterior angle is equal to the sum of the non-adjacent interior angles.
► The sum of the exterior angles of a triangle equal 360 degrees.

Example:

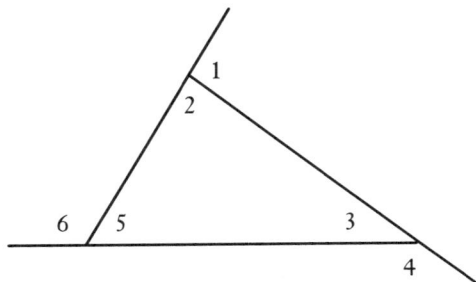

$m\angle 1 + m\angle 2 = 180°$ $m\angle 1 = m\angle 3 + m\angle 5$

$m\angle 3 + m\angle 4 = 180°$ $m\angle 4 = m\angle 2 + m\angle 5$

$m\angle 5 + m\angle 6 = 180°$ $m\angle 6 = m\angle 3 + m\angle 2$

$m\angle 1 + m\angle 4 + m\angle 6 = 360°$

Classifying Triangles

It is possible to classify triangles into three categories based on the number of equal sides:

Scalene Triangle	Isosceles Triangle	Equilateral Triangle
(no equal sides)	(≥ two equal sides)	(all sides equal)

It is also possible to classify triangles into three categories based on the measure of the greatest angle:

Acute Triangle	Right Triangle	Obtuse Triangle
greatest angle is acute	greatest angle is 90°	greatest angle is obtuse

▶ ANGLE-SIDE RELATIONSHIPS

Knowing the angle-side relationships in isosceles, equilateral, and right triangles may be helpful in work related situations.

▶ In isosceles triangles, equal angles are opposite equal sides.

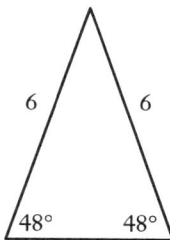

▶ In equilateral triangles, all sides are equal and all angles are equal. The measure of each of the angles in an equilateral triangle is always 60 degrees.

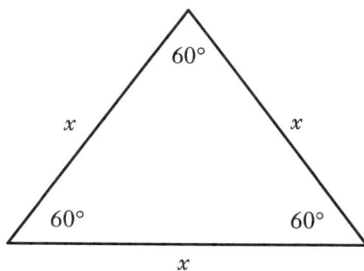

► In a right triangle, the side opposite the right angle is called the **hypotenuse** and the other sides are called **legs**. The box in the angle of the 90-degree angle symbolizes that the triangle is in fact a right triangle.

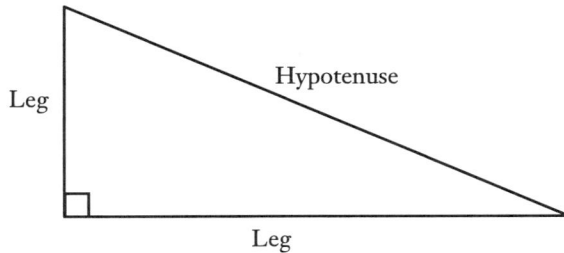

► PYTHAGOREAN THEOREM

The **Pythagorean theorem** is an important tool for working with right triangles. It states: $a^2 + b^2 = c^2$, where a and b represent the legs and c represents the hypotenuse.

This theorem makes it easy to find the length of any side as long as the measure of two sides is known. So, if leg $a = 1$ and leg $b = 2$ in the triangle below, it is possible to find the measure of the hypotenuse, c.

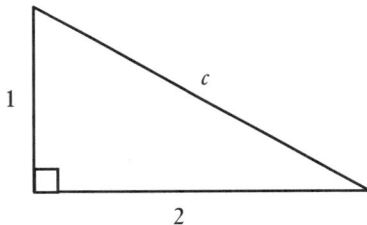

$a^2 + b^2 = c^2$
$1^2 + 2^2 = c^2$
$1 + 4 = c^2$
$5 = c^2$
$\sqrt{5} = c$

► COMPARING TRIANGLES

Triangles are said to be congruent (indicated by the symbol ≅) when they have exactly the same size and shape. Two triangles are congruent if their corresponding parts (their angles and sides) are congruent. Sometimes, it is easy to tell if two triangles are congruent by looking. However, in geometry, it must be able to be proven that the triangles are congruent.

If two triangles are congruent, one of the three criteria listed below must be satisfied.

Side-Side-Side (SSS)	The side measures for both triangles are the same.
Side-Angle-Side (SAS)	Two sides and the angle between them are the same.
Angle-Side-Angle (ASA)	Two angles and the side between them are the same.

▶ POLYGONS

A **polygon** is a closed figure with three or more sides, for example triangles, rectangles, pentagons, etc.

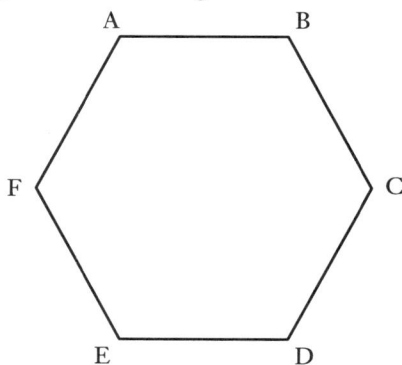

Terms Related to Polygons
  ▶ Vertices are corner points, also called **endpoints**, of a polygon. The vertices in the above polygon are: **A, B, C, D, E,** and **F** and they are always labeled with capital letters.
  ▶ A regular polygon has sides and angles that are all equal.
  ▶ An equiangular polygon has angles that are all equal.

Angles of a Quadrilateral
A **quadrilateral** is a four-sided polygon. Since a quadrilateral can be divided by a diagonal into two triangles, the sum of its interior angles will equal 180 + 180 = 360 degrees.

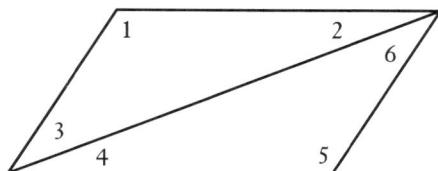

$$m\angle 1 + m\angle 2 + m\angle 3 + m\angle 4 + m\angle 5 + m\angle 6 = 360°$$

Interior Angles

To find the sum of the interior angles of any polygon, use this formula:

$S = 180(x - 2)°$, where x = the number of sides of the polygon.

Example:

Find the sum of the interior angles in the polygon below:

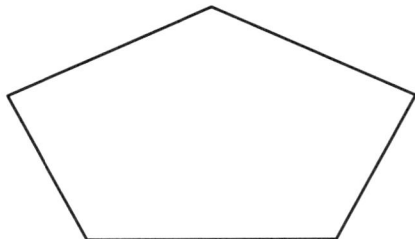

The polygon is a pentagon which has five sides, so substitute 5 for x in the formula:

$S = (5 - 2) \times 180°$

$S = 3 \times 180°$

$S = 540°$

Exterior Angles

Similar to the exterior angles of a triangle, the sum of the exterior angles of any polygon equals 360 degrees.

Similar Polygons

If two polygons are similar, their corresponding angles are equal and the ratios of the corresponding sides are in proportion.

Example:

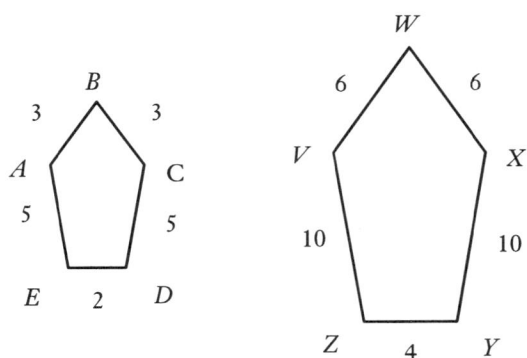

$$\angle A = \angle V = 140°$$
$$\angle B = \angle W = 60°$$
$$\angle C = \angle X = 140°$$
$$\angle D = \angle Y = 100°$$
$$\angle E = \angle Z = 100°$$

$\dfrac{AB}{VW}$	$\dfrac{BC}{WX}$	$\dfrac{CD}{XY}$	$\dfrac{DE}{YZ}$	$\dfrac{EA}{ZV}$
$\dfrac{3}{6}$ =	$\dfrac{3}{6}$ =	$\dfrac{5}{10}$ =	$\dfrac{5}{10}$ =	$\dfrac{2}{4}$

These two polygons are similar because their angles are equal and the ratios of the corresponding sides are in proportion.

▶ PARALLELOGRAMS

A **parallelogram** is a quadrilateral with two pairs of parallel sides.

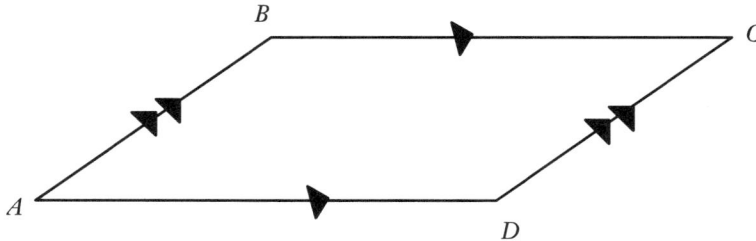

In the figure above, $\overline{AB} \parallel \overline{CD}$ and $\overline{BC} \parallel \overline{AD}$. Parallel lines are symbolized with matching numbers of triangles or arrows.

A parallelogram has:

▶ opposite sides that are equal ($AB = CD$ and $BC = AD$)
▶ opposite angles that are equal ($m\angle A = m\angle C$ and $m\angle B = m\angle D$)
▶ consecutive angles that are supplementary ($m\angle A + m\angle B = 180°$, $m\angle B + m\angle C = 180°$, $m\angle C + m\angle D = 180°$, $m\angle D + m\angle A = 180°$)

Special Types of Parallelograms

▶ A **rectangle** is a parallelogram that has four right angles.

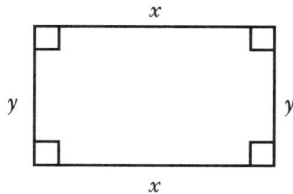

▶ A **rhombus** is a parallelogram that has four equal sides.

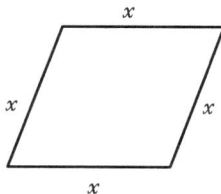

▶ A **square** is a parallelogram in which each of the angles is equal to 90 degrees and all sides are equal to each other. A square is a special case of a rectangle where all the sides are equal. A square is also a special type of rhombus where all the angles are equal.

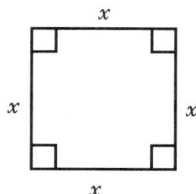

▷ CIRCLES

Remember the following facts about circles:

- ▶ A circle contains 360 degrees.
- ▶ The midpoint of a circle is called the *center*.
- ▶ The distance around a circle is called the *circumference*.
- ▶ The line segment that goes through a circle, with its endpoints on the circle is called a *chord*.
- ▶ A chord that goes directly through the center of a circle—the longest line that can be drawn in a circle—is called the *diameter*.
- ▶ The line from the center of a circle to a point on the circle (half of the diameter) is called the *radius*.
- ▶ The area of a circle is $A = \pi r^2$.
- ▶ The circumference of a circle is $2\pi r$, or πd.

GEOMETRY PRACTICE

Now it's time to practice your skills. Answer the following 15 questions, and then review the answer explanations that follow.

1. Paul needs to carpet the rectangular room below. How many square feet of carpet are needed for wall-to-wall carpeting?

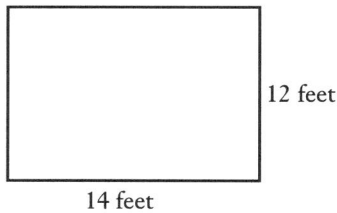

12 feet

14 feet

   a. 26 square feet
   b. 168 square feet
   c. 52 square feet
   d. 40 square feet

2. A bag of concrete mix makes 15 cubic feet of concrete. How many bags of concrete mix need to be purchased to fill a walkway that is 50 cubic feet?
   a. 3
   b. 4
   c. 5
   d. 6

3. A carpenter is using the diagram below to build a room. What is the length of the missing side?

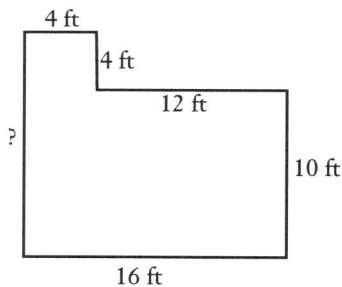

4 ft

4 ft

12 ft

?

10 ft

16 ft

   a. 6 feet
   b. 10 feet
   c. 12 feet
   d. 14 feet

4. A carpenter maintains a home office and uses it as a tax deduction. To determine the amount of his deduction he must calculate the area of the office. If the room that the carpenter uses as an office measures 10 feet by 11 feet, what is the area?

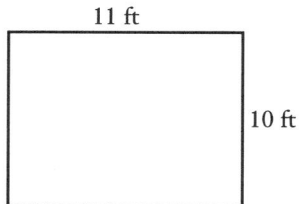

11 ft

10 ft

    a. 100 square feet
    b. 42 square feet
    c. 110 square feet
    d. 84 square feet

5. The perimeter of a rectangular garden is 50 feet. Find the width of the garden if the length is 15 feet.
    a. 10 feet
    b. 5 feet
    c. 20 feet
    d. 7 feet

6. A farmer has 24 feet of fencing. She wants to create the animal pen with the greatest possible area using exactly 24 feet of fencing. Which of the following dimensions would create the pen with the greatest area and use exactly 24 feet of fencing?
    a. 6 feet by 6 feet
    b. 8 feet by 4 feet
    c. 2 feet by 10 feet
    d. 10 feet by 14 feet

7. A mailman walks the route below. How many miles does he walk?

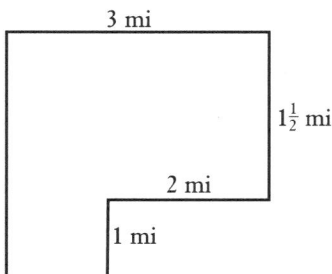

3 mi

$1\frac{1}{2}$ mi

2 mi

1 mi

    a. $7\frac{1}{2}$ miles
    b. 10 miles
    c. 11 miles
    d. $11\frac{1}{2}$ miles

8. A slab of concrete is needed on the third floor of a building. The contractor must calculate the weight of the concrete to ensure that the building will be structurally sound. The concrete slab will be 10 feet by 12 feet and 2 feet thick. The concrete weighs 95 pounds per cubic foot. What is the weight of the concrete slab? (*Volume = length × width × height*)

    a. 240 pounds

    b. 2,950 pounds

    c. 10,600 pounds

    d. 22,800 pounds

9. How much fencing is needed to enclose a circular garden with a diameter of 12 feet? (Use 3.14 for π, and round your answer to the next whole foot.)

    a. 38 feet

    b. 36 feet

    c. 453 feet

    d. 114 feet

10. A rectangular box must be covered with tin sheeting. The box measures 36 inches by 15 inches by 20 inches. What is the total surface area to be covered?

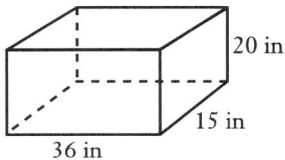

    a. 10,800 square inches

    b. 3,120 square inches

    c. 1,440 square inches

    d. 71 square inches

11. How many cubic feet of concrete are needed to form a circular column seven feet high with a diameter of three feet? (Volume = $\pi r^2 h$; use 3.14 for π, and round your answer to the next cubic foot.)

    a. 200

    b. 198

    c. 21

    d. 50

12. A wire must be strung from the top of a 24-foot pole to a point 18 feet from the base of the pole. How long will the wire be?

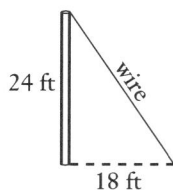

24 ft | *wire* | 18 ft

  a. 42 feet
  b. 60 feet
  c. 30 feet
  d. 25 feet

13. A painter must paint a room that measures 12 feet by 15 feet. The ceiling of the room is 8 feet high. Each gallon of paint costs $25.50 and covers 300 square feet of wall. How much will the paint cost him assuming that the painter will only paint one coat on each wall?
  a. $105.50
  b. $75.50
  c. $65.00
  d. $51.00

14. A circular walkway is installed around a circular pool. The radius of the pool is six feet and the walkway is three feet wide at all points. Find the area of the walkway. (Use 3.14 for π, and round your answer to the next square foot.)

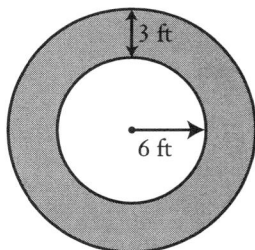

3 ft | 6 ft

  a. 10 square feet
  b. 19 square feet
  c. 109 square feet
  d. 142 square feet

15. The floor plan of a clothing store is shown below. The rent for the store is $3 per square foot. Determine the rent for the clothing store.

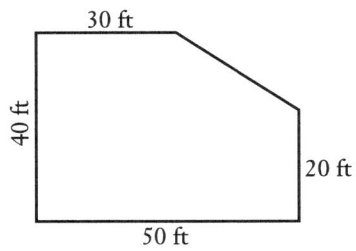

    a. $1,800

    b. $6,000

    c. $5,400

    d. $1,200

ANSWERS

1. **b.** Multiply the length, 14 feet, by the width, 12 feet, to find the area of the rectangle; 14×12 = 168 square feet.

2. **b.** Divide the size of the walkway, 50 cubic feet, by the amount that each bag makes, 15 cubic feet; $50 \div 15 = 3.3$. A little over 3 bags are needed to fill the walkway. Therefore, 4 bags need to be purchased.

3. **d.** The missing side is parallel to the sides that are four feet and ten feet. Those two sides together are the length of the missing side; $4 + 10 = 14$ feet.

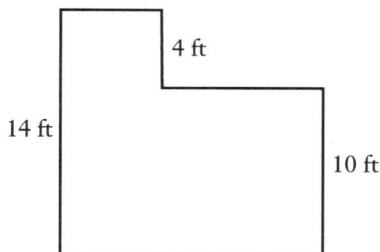

4. **c.** Multiply the length by the width to find the area; $10 \times 11 = 110$ square feet.

5. **a.** The formula for the perimeter of a rectangle is $P = l + l + w + w$; $50 = 15 + 15 + w + w$. Since the two lengths add up to 30, the two widths must add up to 20 to make the perimeter of 50. Since the two widths are the same they must each be 10 feet.

6. **a.** Choice **d** can be eliminated because the dimensions do not make a perimeter of 24 feet ($10 + 10 + 14 + 14 = 48$). The other choices do make a perimeter of 24 feet. Determine which dimensions create the greatest area by multiplying the length by the width; $6 \times 6 = 36$, $8 \times 4 = 32$, and $2 \times 10 = 20$. Therefore, a 6-foot by 6-foot pen has the greatest area.

7. **c.** Find the missing side lengths (shown below) then add all the side lengths; $3 + 1\frac{1}{2} + 2 + 1 + 1 + 2\frac{1}{2} = 11$ miles.

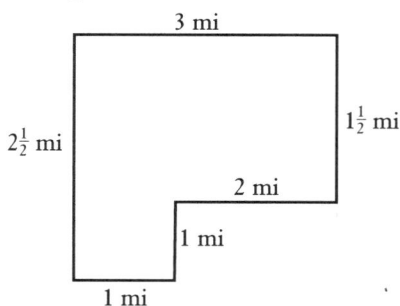

8. **d.** Calculate the number of cubic feet of concrete by finding the volume of concrete; $10 \times 12 \times 2 = 240$ cubic feet. Find the weight of the concrete by multiplying the number of cubic feet by the weight per cubic foot; $240 \times 95 = 22,800$ pounds.

9. **a.** The formula for the circumference of a circle is $C = \pi d$. Therefore, the circumference of the circular garden is $(3.14)(12) = 37.68$ feet. The question asks for the answer to the next whole foot which is 38 feet.

10. **b.** Find the area of each of the six sides. Two of the sides (front and back) are 36 feet by 20 feet and have an area of 720 square feet each. The top and bottom are 36 feet by 15 feet and have an area of 540 square feet each. The left and right sides are 15 feet by 20 feet and have an area of 300 square feet each. Find the total surface area by adding up the areas of all six sides; 720 + 720 + 540 + 540 + 300 + 300 = 3,120 square feet.

11. **d.** The column is a cylindrical shape. To find the volume of the cylinder, multiply the area of the base (a circle) by the height of the column. The area of the base is $(3.14)(1.5)^2 = 7.065$ square feet. Multiply this area by the height to find the volume; $7.065 \times 7 = 49.455$ cubic feet. The question asks for the answer to the next cubic foot, 50 cubic feet.

12. **c.** The pole, wire, and ground form a right triangle. Therefore, the Pythagorean theorem can be used to find the missing side of the triangle. The Pythagorean theorem is $a^2 + b^2 = c^2$, where c is the longest side of the triangle, the hypotenuse. In this example, the hypotenuse is the wire. Substitute the values into the equation and solve for c.

$$24^2 + 18^2 = c^2$$
$$576 + 324 = c^2$$
$$900 = c^2$$
$$\sqrt{900} = \sqrt{c^2}$$
$$30 = c$$

The length of the wire is 30 feet.

13. **d.** Find the area of the walls. Two walls are 15 feet by 8 feet and the other two walls are 12 feet by 8 feet; $15 \times 8 = 120$ square feet; $12 \times 8 = 96$ square feet. The total area of the walls in the room is 120 + 120 + 96 + 96 = 432 square feet. Next, find the number of gallons of paint that must be purchased. Each gallon covers 300 square feet, so two gallons will be needed. Last, calculate the cost of those two gallons of paint; $2 \times \$25.50 = \$51.00$.

14. **d.** Find the area of the entire circle and subtract the area of the pool. The radius of the entire circle is nine feet because six feet plus three feet is nine feet. The area of the entire circle is $(9^2)(3.14) = 254.34$ square feet. The area of the pool is $(6^2)(3.14) = 113.04$ square feet. Subtract the area of the pool from the area of the entire circle to find the area of the walkway; $254.34 - 113.04 = 141.3$ square feet. The question asks for the answer to the next square foot, 142 square feet.

15. **c.** Divide the floorplan into two rectangles and a triangle as shown below. Find the area of each shape, then add the areas together. The area of the rectangle on the left is $30 \times 40 = 1,200$ square feet. The area of the rectangle on the right is $20 \times 20 = 400$ square feet. The area of the triangle is $\frac{1}{2} \times 20 \times 20 = 200$ square feet. To find the area of the entire figure, add the three areas together 1,200 + 400 + 200 = 1,800 square feet. The rent is $3 per square foot. Therefore the total rent is $1,800 \times \$3 = \$5,400$.

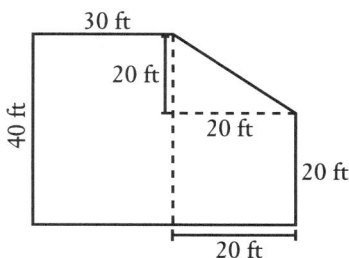

CHAPTER 7

Word Problem and Data Analysis Review

Many people struggle with word problems and graphs. In this chapter, you will learn how to set up and solve word problems, and understand graphs, charts, tables, and diagrams with confidence.

▶ TRANSLATING WORDS INTO NUMBERS

The most important skill needed for word problems is being able to translate words into mathematical operations. The following will be helpful in achieving this goal by providing common examples of English phrases and their mathematical equivalents.

"Increase" means add.
Example:
A number increased by five = $x + 5$.

"Less than" means subtract.
Example:
Ten less than a number = $x - 10$.

"Times" or "product" means multiply
Example:
Three times a number = $3x$.

"Times the sum" means to multiply a number by a quantity.
Example:
Five times the sum of a number and three = $5(x + 3)$.

Two variables are sometimes used together.
Example:
A number y exceeds five times a number x by ten.
$y = 5x + 10$

"Of" means multiply.
Example:
10% of 100 = $10\% \times 100$

"Is" means equals.
Example:
15 is 14 plus 1 becomes 15 = 14 + 1.

▶ ASSIGNING VARIABLES IN WORD PROBLEMS

It may be necessary to create and assign variables in a word problem. To do this, first identify an unknown and a known. The known may not be a specific numerical value, but the problem should indicate something about its value.

Examples:
1. Max has worked for three more years than Ricky.
Unknown: Ricky's work experience = x
Known: Max's experience is three more years = $x + 3$
Therefore, Ricky's experience = x and Max's experience = $x + 3$

2. Heidi made twice as many sales as Rebecca.

Unknown: number of sales Rebecca made = x

Known: number of sales Heidi made is twice Rebecca's amount = $2x$

3. There are six less than four times the number of pens than pencils.

Unknown: the number of pencils = x

Known: the number of pens is six less than four times the number of pencils = $4x - 6$

▶ RATIO

A **ratio** is a comparison of two quantities measured in the same units. It is symbolized by the use of a colon—$x : y$. Ratios can also be expressed as fractions ($\frac{x}{y}$) or using words (x to y).

Ratio problems are solved using the concept of multiples.

Example:

A bag contains 60 screws and nails. The ratio of the number of screws to nails is 7 : 8. How many of each kind are there in the bag?

Solution:

From the problem, it is known that 7 and 8 share a multiple and that the sum of their product is 60. Write and solve the following equation.

$$7x + 8x = 60$$
$$\frac{15x}{15} = \frac{60}{15}$$
$$x = 4$$

Therefore, there are (7)(4) = 28 screws and (8)(4) = 32 nails.

Check: 28 + 32 = 60 screws, $\frac{28}{32} = \frac{7}{8}$.

▶ VARIATION

Variation is a term referring to a constant ratio in the change of a quantity.

▶ A quantity is said to vary directly with another if they both change in an equal direction. In other words, two quantities vary directly if an increase in one causes an increase in the other. This is also true if a decrease in one causes a decrease in the other. The ratio, however, must be the same.

Example:

If it takes a total of 58.5 hours to train 300 new employees, how many hours of training will it take for 800 employees?

Solution:

Since each employee needs about the same amount of training, you know that they vary directly. Therefore, you can set the problem up the following way.

$\frac{employees}{hours} \rightarrow \frac{300}{58.5} = \frac{800}{x}$

Cross multiply to solve.

$(800)(58.5) = 300x$

$\frac{46,800}{300} = \frac{300x}{300}$

$156 = x$

Therefore, it would take 156 hours to train 800 employees.

▶ If two quantities change in opposite directions, they are said to vary inversely. This means that as one quantity increases, the other decreases, or as one decreases, the other increases.

Example:

If two people plant a field in six days, how may days will it take six people to plant the same field? (Assume each person is working at the same rate.)

Solution:

As the number of people who are planting increases, the days needed to plant decreases. Therefore, the relationship between the number of people and days varies inversely. Because the field remains constant, the two products can be set equal to each other.

2 people × 6 days = 6 people × x days

$2 \times 6 = 6x$

$\frac{12}{6} = \frac{6x}{6}$

$2 = x$

Thus, it would take six people two days to plant the same field.

▶ RATE PROBLEMS

In general, there are three different types of rate problems likely to be encountered in the workplace: cost, movement, and work-output. **Rate** is defined as a comparison of two quantities with different units of measure.

$\text{Rate} = \frac{x \text{ units}}{y \text{ units}}$

Examples:

$$\frac{dollars}{hour}, \frac{cost}{pound}, \frac{miles}{hour}$$

Cost Per Unit

Some problems will require the calculation of unit cost.

Example:

If 100 square feet cost $1,000, how much does 1 square foot cost?

Solution:

$$\frac{\text{Total cost}}{\text{\# of square feet}} = \frac{1,000}{100 \text{ ft}^2}$$

$$= \$10 \text{ per square foot}$$

Movement

In working with movement problems, it is important to use the following formula:

$(Rate)(Time) = Distance$

Example:

A courier traveling at 15 mph traveled from his base to a company in $\frac{1}{4}$ of an hour less than it took when the courier traveled 12 mph. How far away was his drop off?

Solution:

First, write what is known and unknown.

Unknown: time for courier traveling 12 mph = x

Known: time for courier traveling 15 mph = $x - \frac{1}{4}$

Then, use the formula $(Rate)(Time) = Distance$ to find expressions for the distance traveled at each rate:

12 mph for x hours = a distance of $12x$ miles

15 miles per hour for $x - \frac{1}{4}$ hours = a distance of $15x - \frac{15}{4}$ miles.

The distance traveled is the same, therefore, make the two expressions equal to each other:

$$12x = 15x - 3.75$$
$$-15x = -15x$$
$$\frac{-3x}{-3} = \frac{-3.75}{-3}$$
$$x = 1.25$$

Be careful, 1.25 is not the distance; it is the time. Now you must plug the time into the formula $(Rate)(Time) = Distance$. Either rate can be used.

$$12x = \text{distance}$$
$$12(1.25) = \text{distance}$$
$$15 \text{ miles} = \text{distance}$$

Work-Output Problems

Work-Output problems are word problems that deal with the rate of work. The following formula can be used on these problems.

(Rate of work)(time worked) = job or part of job completed

Example:

Danette can wash and wax two cars in six hours, and Judy can wash and wax the same two cars in four hours. If Danette and Judy work together, how long will it take to wash and wax one car?

Solution:

Since Danette can wash and wax 2 cars in 6 hours, her rate of work is 2 cars/6 hours, or one car every three hours. Judy's rate of work is therefore, 2 cars/4 hours, or one car every two hours. In this problem, making a chart will help:

	RATE	TIME	=	PART OF JOB COMPLETED
Danette	$\frac{1}{3}$	x	=	$\frac{1}{3}x$
Judy	$\frac{1}{2}$	x	=	$\frac{1}{2}x$

Since they are both working on only one car, you can set the equation equal to one:
Danette's part + Judy's part = one car.
$\frac{1}{3}x + \frac{1}{2}x = 1$
Solve by using 6 as the LCD for 3 and 2 and clear the fractions by multiplying by the LCD:

$$6(\tfrac{1}{3}x) + 6(\tfrac{1}{2}x) = 6(1)$$
$$2x + 3x = 6$$
$$\frac{5x}{5} = \frac{6}{5}$$
$$x = 1\tfrac{1}{5}$$

Thus, it will take Judy and Danette $1\frac{1}{5}$ hours to wash and wax one car.

▶ DATA ANALYSIS

Data analysis simply means reading graphs, tables, and other graphical forms. You should be able to:

▶ read and understand scatter plots, graphs, tables, diagrams, charts, figures, etc.
▶ interpret scatter plots, graphs, tables, diagrams, charts, figures, etc.
▶ compare and interpret information presented in scatter plots, graphs, tables, diagrams, charts, figures, etc.
▶ draw conclusions about the information provided
▶ make predictions about the data

It is important to read tables, charts, and graphs very carefully. Read all of the information presented, paying special attention to headings and units of measure. This section will cover tables and graphs. The most common types of graphs are scatter plots, bar graphs, and pie graphs. What follows is an explanation of each, with examples for practice.

Tables

All **tables** are composed of **rows** (horizontal) and **columns** (vertical). Entries in a single row of a table usually have something in common, and so do entries in a single column. Look at the table below that shows how many cars, both new and used, were sold during the particular months.

MONTH	NEW CARS	USED CARS
June	125	65
July	155	80
August	190	100
September	220	115
October	265	140

Tables are very concise ways to convey important information without wasting time and space. Just imagine how many lines of text would be needed to convey the same information. With the table, however, it is easy to refer to a given month and quickly know how many total cars were sold. It would also be easy to compare month to month. In fact, practice by comparing the total sales of July with October.

In order to do this, first find out how many cars were sold in each month. There were 235 cars sold in July (155 + 80 = 235) and 405 cars sold in October (265 + 140 = 405). With a little bit of quick arithmetic it can quickly be determined that 170 more cars were sold during October (405 − 235 = 170).

Scatter Plots

Whenever a variable depends continuously on another variable, this dependence can be visually represented in a **scatter plot**. A scatter plot consists of the horizontal (x) axis, the vertical (y) axis, and collected data points for variable y, measured at variable x. The variable points are often connected with a line or a curve. A graph often contains a legend, especially if there is more than one data set or more than one variable. A **legend** is a key for interpreting the graph. Much like a legend on a map lists the symbols used to label an interstate highway, a railroad line, or a city, a legend for a graph lists the symbols used to label a particular data set. Look at the sample graph on the following page. The essential elements of the graph—the x- and y-axis—are labeled. The legend to the right of the graph shows that diamonds are used to represent the variable points in data set 1, while squares are used to represent the variable points in data set 2. If only one data set exists, the use of a legend is not essential.

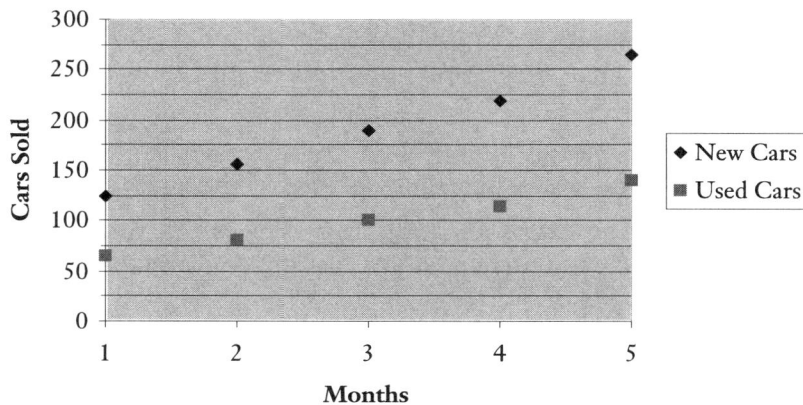

(Note: this is the same data that was used in the previous example for tables.)

The *x*-axis represents the months after new management and promotions were introduced at an automobile dealership. The *y*-axis represents the number of cars sold in the particular month after the changes were made. The diamonds reflect the New Cars sold and the squares show the number of Used Cars sold. What conclusions can be drawn about the sales? Note that the New and Used car sales are both increasing each month at a rather steady rate. The graph also shows that New Cars increase at a higher rate and that there are many more New Cars sold per month.

Try to look for scatter plots with different trends—including:

▶ increase
▶ decrease
▶ rapid increase, followed by leveling off
▶ slow increase, followed by rapid increase
▶ rise to a maximum, followed by a decrease
▶ rapid decrease, followed by leveling off (as in the wavelength example)
▶ slow decrease, followed by rapid decrease
▶ decrease to a minimum, followed by a rise
▶ predictable fluctuation (periodic change, such as a light wave)
▶ random fluctuation (irregular change)

Bar Graphs

Bar graphs are similar to scatter plots. Both have a variable *y* plotted against a variable *x*. However, in bar graphs, data is represented by bars, rather than by points connected with a line. Bar graphs are often used to indicate an amount or level, as opposed to a continuous change. Consider the following bar graph. It illustrates the number of employees who were absent due to illness during a particular week in two different age groups.

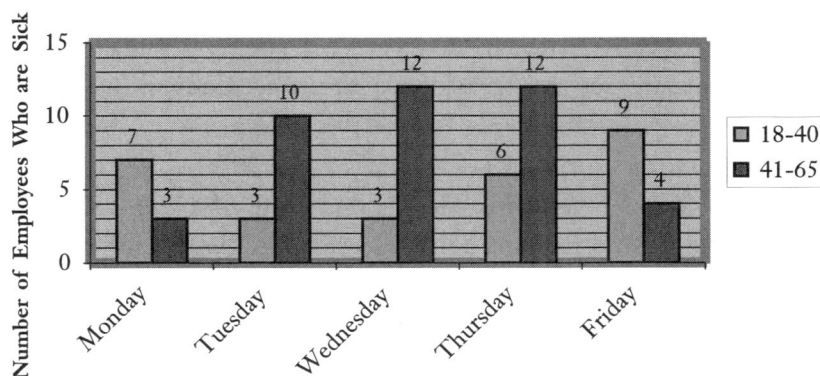

It can be immediately seen that younger employees are sick before and after the weekend. There is also an inconsistent trend for the younger employees with data ranging all over the place. During mid-week the older crowd tends to stay home more often.

How many people on average are sick in the 41–65 age group? To find the average you first must find out how many illnesses occur each week in the particular age group. There are a total of 41 illnesses for a five-day period (3 + 10 + 12 + 12 + 4 = 41). To calculate the average, just divide the total illnesses by the number of days for a total of 8.2 illnesses ($\frac{41}{5}$ = 8.2) or more realistically, 8 absences per day.

Pie Charts and Circle Graphs

Pie graphs are often used to show what percent of a total is taken up by different components of that whole. This type of graph is representative of a whole and is usually divided into percentages. Each section of the chart represents a portion of the whole, and all of these sections added together will equal 100% of the whole. The chart below shows the three styles of model homes in a new development and what percentage of each there is.

Models of Homes

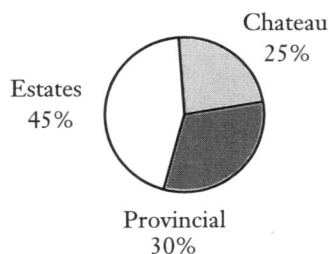

The chart shows the different models of home. Find the percentage of Estates homes. In order to find this percentage look at the pie chart. The categories add up to 100% (25 + 30 + 45 = 100). From the actual chart you can visually see that 45% of the homes are done in the Estates model.

Broken Line Graphs

Broken-line graphs illustrate a measurable change over time. If a line is slanted up, it represents an increase, whereas a line sloping down represents a decrease. A flat line indicates no change.

In the broken line graph below, the number of delinquent payments is charted for the first quarter of the year. Each week the number of outstanding bills is summed and recorded.

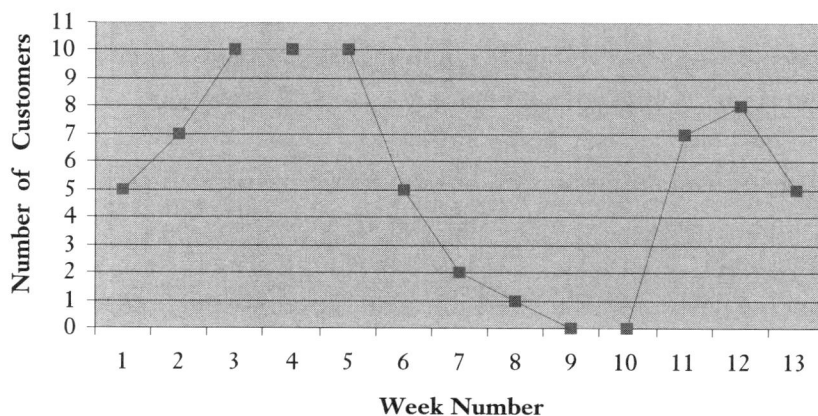

There is an increase in delinquency for the first two weeks and then it maintains for an additional two weeks. There is a steep decrease after week 5 (initially) until the ninth week, where it levels off again but this time at 0. The eleventh week shows a radical increase followed by a little jump up at week 12, and then a decrease to week 13. It is also interesting to see that the first and last weeks have identical values.

Diagrams

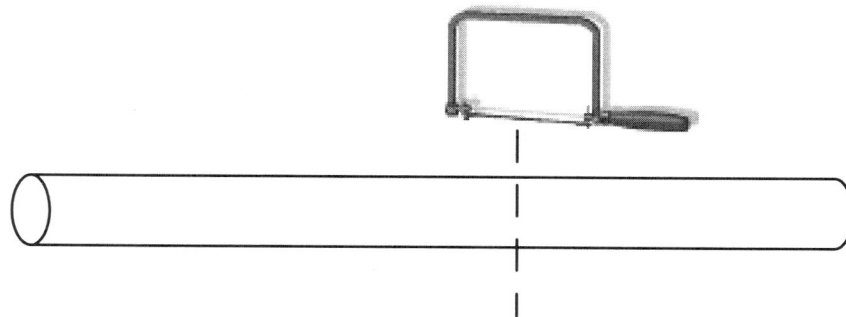

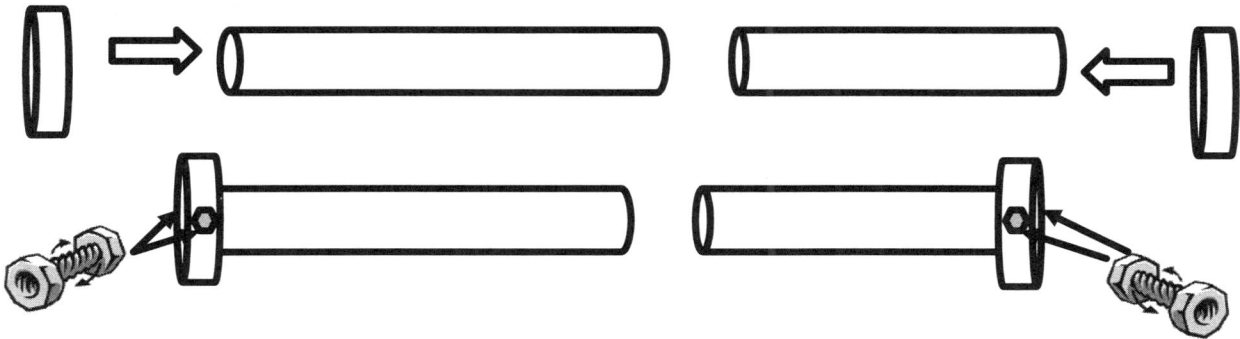

The two-part diagram on the previous page and above shows a sequence of events to construct two new objects out of one pipe and a few other parts. First the instructions show that the pipe must be cut into two pieces with a saw. The next two levels show how the assembly will take place, first adding the end pieces and then bolting in those pieces.

Diagrams could be used to show a sequence of events, a process, an idea, or the relationship between different events or people. Here are some examples that might be found in the workplace:

▶ diagrams showing design elements and construction instructions
▶ diagrams showing management system relationships
▶ diagrams showing correct procedure

When you see a diagram, first ask what the purpose of it is. What is it trying to illustrate? Then look at the different labeled parts of the diagram. What is the function of each part? How are they interrelated? Take a look at the following diagram:

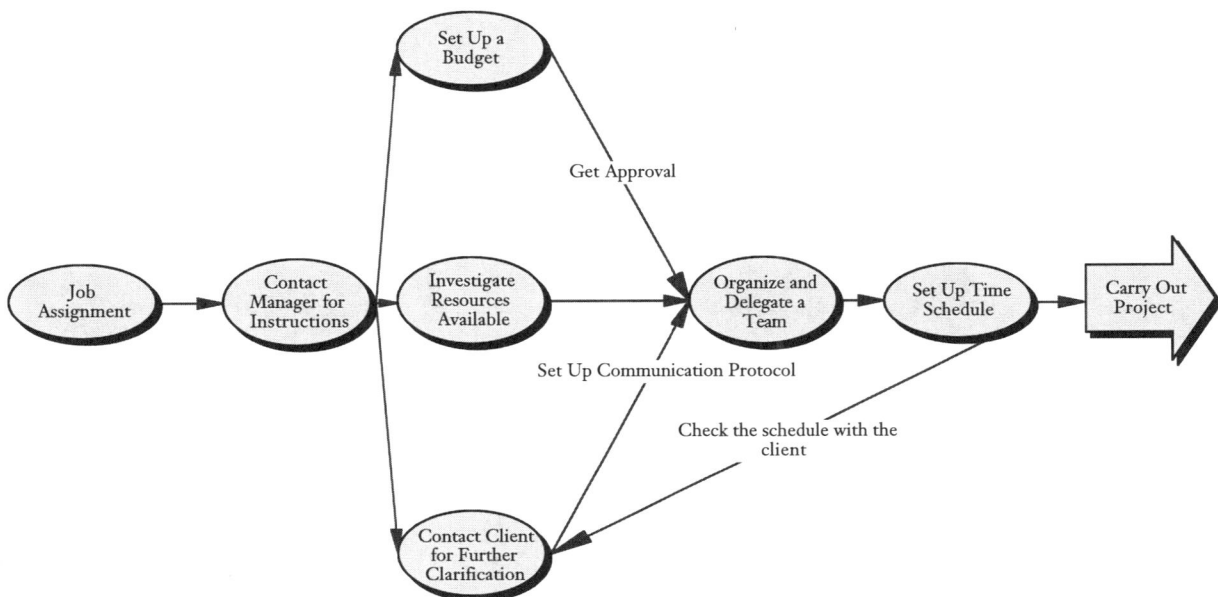

This diagram is a typical chart of how to start a new project. It starts (on the left), by learning about the assignment from the manager and then investigating several aspects of heading a project, including the client, resources, and budget. Once an overall picture is achieved you will then know how many people will be required for the project in order to create an accurate schedule for the project. There are also reminders and further protocols within the diagram in the links.

WORD PROBLEM AND DATA ANALYSIS PRACTICE

Now it's time to practice your skills. Answer the following 15 questions, and then review the answer explanations that follow.

1. The purchaser for the clothing department of a major retailer can purchase 120 women's sweaters at a cost of $900. What is the cost per sweater?
  a. $1,020.00
  b. $13.00
  c. $7.50
  d. $8.00

2. Two pipes are being used to fill a pool. One fills the pool at the rate of five gallons per minute. The other works at the rate of six gallons per minute. How long will the pipes take to fill a 6,000-gallon pool if they are used together? Round your answer to the nearest minute.
  a. 500 minutes
  b. 545 minutes
  c. 1,000 minutes
  d. 1,200 minutes

3. A stockperson at a sporting goods store noted the following sales of balls for the month of January:

Ball Sales for January

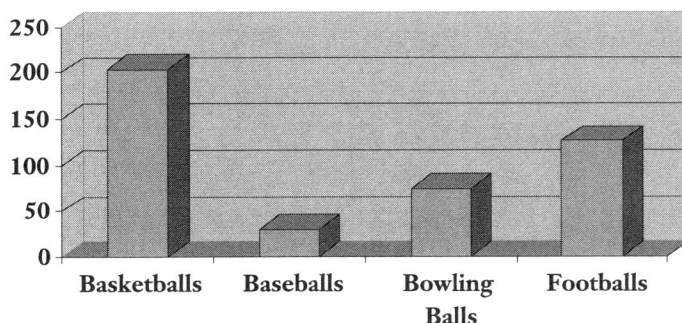

Which of the following statements is most accurate?
  a. The number of basketballs sold was twice the number of footballs sold.
  b. The number of bowling balls sold was half the number of footballs sold.
  c. Three times the number of baseballs sold is equal to the number of bowling balls sold.
  d. The total number of baseballs, bowling balls, and footballs is less than the number of basketballs.

4. The cook at Pizza Place surveyed his customers regarding the type of pizza crust they prefer. The graph of his conclusions follows:

Pizza Crust Preference

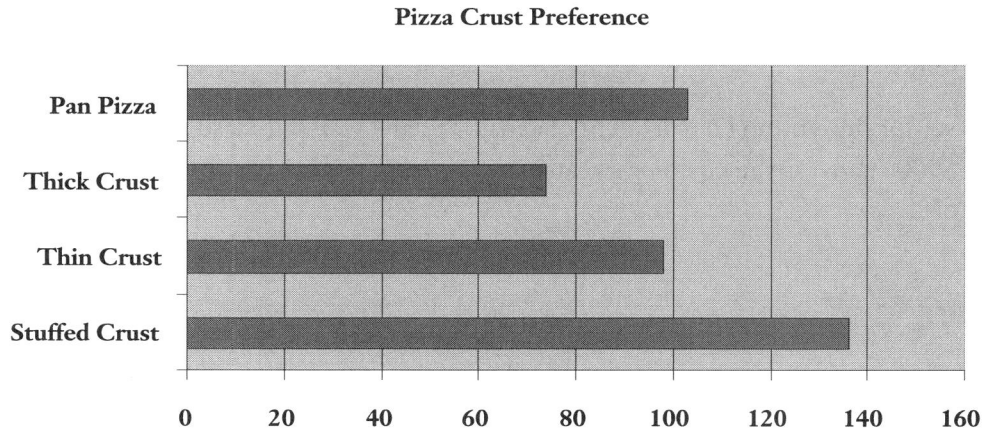

About how many more people prefer pan pizza than thin crust pizza?

a. 5
b. 29
c. 33
d. 24

Refer to the following line graph to answer questions 5 and 6.

Number of Days Weather Permits Outdoor Construction Work

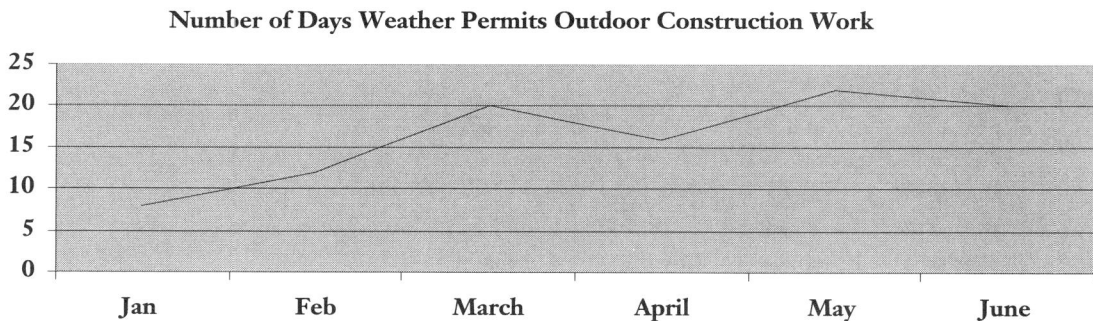

5. In which month can workers work outside the most?

a. March
b. May
c. April
d. June

6. If an average worker gets paid $15.50/hour working construction, and works a 10-hour day, about how much money can he expect to make working outside in January?
   a. $155.00
   b. $124.00
   c. $1,240.00
   d. $1,860.00

Refer to the following table to answer questions 7–9.

HOURLY WAGES OF EMPLOYEES AT BARB'S BAR-B-QUE			
	YEAR 1	**YEAR 2**	**YEAR 3**
Cashier	$7.20	$7.50	$7.90
Hostess	$6.85	$7.00	$7.30
Waitstaff	$8.10	$8.80	$9.75
Cook	$9.20	$9.95	$10.40

7. What is the hourly wage of a second-year waitperson at Barb's?
   a. $8.10
   b. $8.80
   c. $9.95
   d. $9.75

8. If a third year cook works 40 hours per week, how much can he expect to earn?
   a. $398 per week
   b. $390 per week
   c. $368 per week
   d. $416 per week

9. What is the difference in the weekly paycheck of a first-year Cashier and a second-year Hostess, assuming each works a 35-hour week?
   a. $7.00
   b. $245.00
   c. $252.00
   d. $497.00

Refer to the following graph to answer questions 10–11. The graph represents sales data for a cell phone distributor for the months since the company was created.

Record of 2003 Sales Data

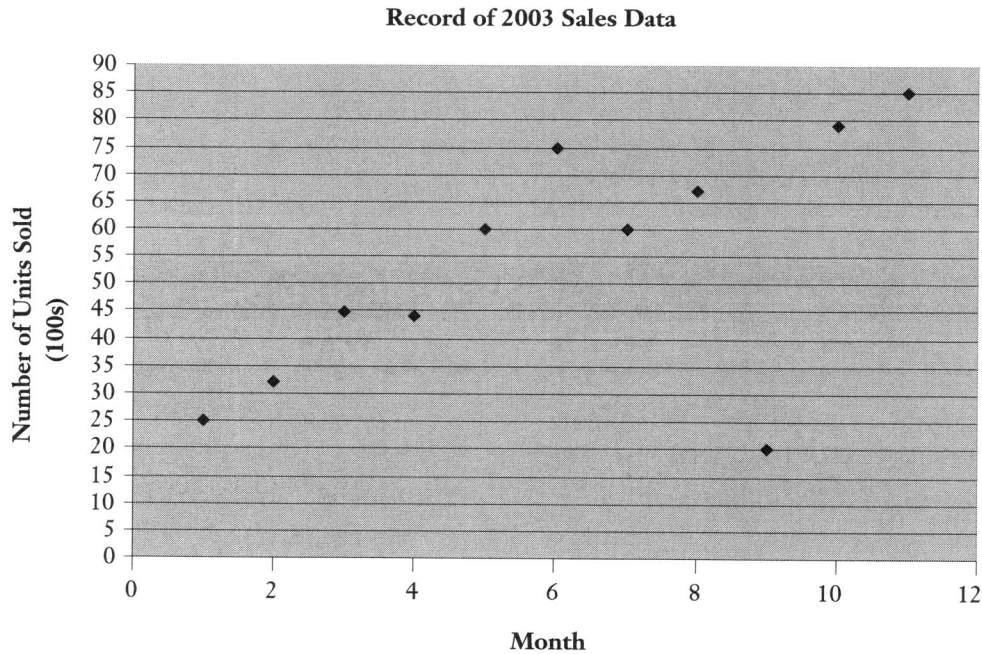

10. Which is larger, the percent increase between months 1 and 2, or the percent decrease between months 6 and 7?
  a. The percent increase between months 1 and 2 is greater.
  b. The percent decrease between months 6 and 7 is greater.
  c. They are the same.
  d. The answer cannot be determined using the information provided.

11. Your boss asks you to write a report describing the general trend of sales this year. Based on the graph, which sentence below most accurately describes the rate of change in sales in 2003?
  a. Sales in 2003 only increased.
  b. Sales in 2003 showed no pattern, they were completely random from month to month.
  c. Sales in 2003 started low, but never dropped beneath the original amounts in month 1.
  d. Sales in 2003 generally showed an increase, although some months showed a decrease.

12. The developer for a new subdivision has collected the following data from previous building projects. If he plans to build 50 homes in the new subdivision, how many should be painted white if he is to remain consistent with what purchasers have chosen in the past?

Color Preferene for Home Buyers

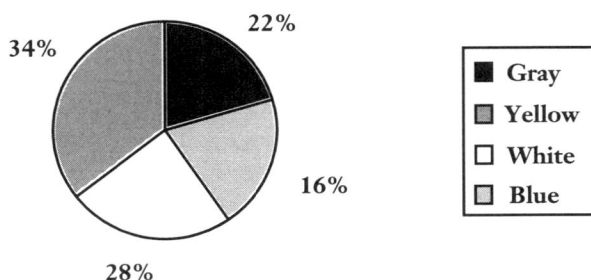

a. 8
b. 17
c. 11
d. 14

13. A housecleaner has run out of window cleaning fluid and wishes to purchase more. At the store, she finds that a 44-ounce bottle of cleaner costs $2.75, while a 55-ounce bottle of a different brand costs $3.70. Which is the better buy?
a. the 44-ounce bottle
b. the 55-ounce bottle
c. The costs are the same per unit.
d. The answer cannot be determined from the information provided.

14. Joe needs to deliver a washing machine across town. The company truck can travel 65 mph on the highway, which comprises 13 miles of his trip. For the remaining 12 miles, Joe will average 24 mph. Assuming he does not hit traffic, at what time must Joe leave the warehouse in order to arrive at his destination by 1:30 P.M.?
a. 1:00 P.M.
b. 12:55 P.M.
c. 12:50 P.M.
d. 12:48 P.M.

15. Whiz-Bang Construction can construct a 1,400-square-foot home for $20,000. Smithco Building can construct a 2,200-square-foot home for $28,000. DJK Builders can construct a 2,500-square-foot home for $35,000. If a contractor is looking for a construction company to construct 2,800-square-foot homes in a new subdivision, and uses the most economical of the three construction companies, approximately how much will the construction of each home in the subdivision cost? Round your answer to the nearest thousand dollars.

    a. $36,000

    b. $35,000

    c. $40,000

    d. $39,000

ANSWERS

1. c. To find the cost per sweater, divide the total cost ($900) by the number of sweaters (120); $900 divided by 120 = $7.50.

2. b. With both pipes working, the pool is being filled at a rate of 11 gallons per minute ($\frac{5 \text{ gallons}}{\text{minute}} + \frac{6 \text{ gallons}}{\text{minute}} = \frac{11 \text{ gallons}}{\text{minute}}$). Now, divide the total amount of gallons in the pool (6,000) by the rate ($\frac{11 \text{ gallons}}{\text{minute}}$); 6,000 divided by 11 = 545.45. Round to the nearest minute: 545 minutes.

3. c. The number of baseballs sold is approximately 25 (bar is halfway between 0 and 50). The number of bowling balls sold is approximately 75 (bar is halfway between 50 and 100). Three times 25 is equal to 75.

4. a. A little more than 100 people prefer pan pizza, while a little fewer than 100 people prefer thin crust pizza. The difference between these numbers is closest to 5.

5. b. Since the point that corresponds to the month of May is the highest, this is the month during which workers are able to work outside the most.

6. c. This is a multi-step problem. First, calculate how much workers will make for a 10-hour day. At $15.50 per hour, the average worker will make $155.00 per day working 10-hour days ($15.50 times 10). Since the graph shows that work can be done outside on approximately eight days in January, the worker can expect to make $155.00 every day, so $155.00 times 8 = $1,240.

7. b. Look across the row (horizontal) labeled *Waitstaff*. Find the corresponding column (vertical) for Year 2. Find the value that corresponds to the location where the column and the row intersect. A second-year waitperson would earn $8.80 an hour at Barb's Bar-B-Que.

8. d. The hourly pay of a third year cook is $10.40 (fourth row, third column). If the cook works 40 hours, he or she will earn 40 times $10.40 = $416 per week.

9. a. A first year cashier would earn $7.20 per hour for 35 hours; $7.20 times 35 = $252. A second year hostess would earn $7.00 per hour for 35 hours; $7.00 times 35 = $245. To find the difference, subtract: $252 − $245 = $7.00. A first year cashier would earn $7.00 more than a second year hostess.

10. a. To calculate the percent increase between months 1 and 2, you must look at the graph. Month 2 reads sales of about 3,200 and month 1 reads sales of 2,500. To calculate the percent increase, you must find the difference between the two numbers, then divide the difference by the **original** amount. Subtract first: 3,200 − 2,500 = 700. Now, you must determine what percent 700 is of 2,500, since 2,500 is the amount of sales in the first of the two months; 700 is x% of 2,500. To calculate percent, divide 700 by 2,500. The answer is 0.28, which is 28%. Now you must calculate the percent decrease between months 6 and 7. By reading the graph, you see that sales in month 6 were 7,500 and the sales in month 7 were 6,000. Use the same process to calculate the percent decrease; 7,500 − 6,000 = 1,500; 1,500 is x% of 7,500; 1,500 divided by 7,500 = .20 or 20%. The answer is that the percent increase from month 1 to month 2 is greater than the percent decrease from month 6 to month 7.

11. d. Start by comparing the information on the graph to the statements in each of the answer choices. Choice **a** states that sales in 2003 only increased. However, if you review months 4, 7, and 9, you will see that sales decrease from their respective previous month's sales. Choice **b** states that sales in 2003 showed no pattern, they were completely random month to month. If you look at the graph, you will see that the general (overall) pattern of the graph shows an increase in sales. Although the sales figures vary, it is not accurate to say they are completely random. Choice **c** states that sales in 2003 started low, but never dropped beneath the original amounts in month 1. If you look at the graph, specifically at month 9, you will see that although month 1's sales were 2,500, month 9's sales were 2,000. This makes this answer choice incorrect. Finally, choice **d** states that sales in 2003 generally showed an increase, although some months showed a decrease. This statement accurate describes the rate of change during the 12 months of 2003.

12. **d.** 28% of people chose white as the color of their home in his previous experience. Since 50 homes are to be constructed in the new neighborhood, 28% of this number should be painted white; 28% of 50 (.28 times 50) = 14; 14 homes should be painted white.

13. **a.** The first cleaner costs $2.75 for 44 ounces. Since the cost per ounce is needed, divide $2.75 by 44 to get about $.0625 per ounce, or about $.06 per ounce. The second cleaner costs $3.70 for 55 ounces; $3.70 divided by 55 = $.067 per ounce, or about $.07 per ounce. Therefore, the first cleaner is the better buy because it is cheaper per ounce.

14. **d.** For the highway portion of the trip, Joe will average 65 mph for 13 miles. Since 13 is $\frac{1}{5}$ of 65, this portion of the trip will take Joe $\frac{1}{5}$ of an hour, or 12 minutes. For the remaining 12 miles of the trip, Joe will average 24 mph. Since 12 is half of 24, this portion of the trip will take half an hour or 30 minutes. The total time will be 12 minutes + 30 minutes = 42 minutes. To arrive by 1:30 P.M., Joe must leave no later than 12:48 P.M. since 1:30 P.M. – 42 minutes = 12:48 P.M. Therefore, the correct answer is 12:48 P.M.

15. **a.** The 1,400-square-foot home costs about $14.29 per square foot: $20,000 divided by 1,400 square feet = $14.29 per square foot. The 2,200-square-foot home costs about $12.73 per square foot: $28,000 divided by 2,200 square feet = $12.73 per square foot. The 2,500-square-foot home costs about $14 per square foot: $35,000 divided by 2,500 square feet = $14 per square foot. The cheapest price per square foot is $12.73 by Smithco. If this company is hired to construct 2,800-square-foot homes, it will cost $35,644 per home: 2,800 square feet times $12.73 per square foot = $35,644. Rounded to the nearest thousand, the answer is $36,000.

8

Post-test

This post-test was designed to show you how well you learned the material presented in *Math for the Trades*. The questions on this test are similar to those found in the pretest, so you can compare your results both before and after completing the review lessons and practice questions in this book.

When you have answered the following 100 questions, take time to review the answer explanations. If there is anything that still does not make sense, go back to the chapters and review the necessary material.

Good luck!

▶ ANSWER SHEET

1.	ⓐ ⓑ ⓒ ⓓ	34.	ⓐ ⓑ ⓒ ⓓ	68.	ⓐ ⓑ ⓒ ⓓ
2.	ⓐ ⓑ ⓒ ⓓ	35.	ⓐ ⓑ ⓒ ⓓ	69.	ⓐ ⓑ ⓒ ⓓ
3.	ⓐ ⓑ ⓒ ⓓ	36.	ⓐ ⓑ ⓒ ⓓ	70.	ⓐ ⓑ ⓒ ⓓ
4.	ⓐ ⓑ ⓒ ⓓ	37.	ⓐ ⓑ ⓒ ⓓ	71.	ⓐ ⓑ ⓒ ⓓ
5.	ⓐ ⓑ ⓒ ⓓ	38.	ⓐ ⓑ ⓒ ⓓ	72.	ⓐ ⓑ ⓒ ⓓ
6.	ⓐ ⓑ ⓒ ⓓ	39.	ⓐ ⓑ ⓒ ⓓ	73.	ⓐ ⓑ ⓒ ⓓ
7.	ⓐ ⓑ ⓒ ⓓ	40.	ⓐ ⓑ ⓒ ⓓ	74.	ⓐ ⓑ ⓒ ⓓ
8.	ⓐ ⓑ ⓒ ⓓ	41.	ⓐ ⓑ ⓒ ⓓ	75.	ⓐ ⓑ ⓒ ⓓ
9.	ⓐ ⓑ ⓒ ⓓ	42.	ⓐ ⓑ ⓒ ⓓ	76.	ⓐ ⓑ ⓒ ⓓ
10.	ⓐ ⓑ ⓒ ⓓ	43.	ⓐ ⓑ ⓒ ⓓ	77.	ⓐ ⓑ ⓒ ⓓ
11.	ⓐ ⓑ ⓒ ⓓ	44.	ⓐ ⓑ ⓒ ⓓ	78.	ⓐ ⓑ ⓒ ⓓ
12.	ⓐ ⓑ ⓒ ⓓ	45.	ⓐ ⓑ ⓒ ⓓ	79.	ⓐ ⓑ ⓒ ⓓ
13.	ⓐ ⓑ ⓒ ⓓ	46.	ⓐ ⓑ ⓒ ⓓ	80.	ⓐ ⓑ ⓒ ⓓ
14.	ⓐ ⓑ ⓒ ⓓ	47.	ⓐ ⓑ ⓒ ⓓ	81.	ⓐ ⓑ ⓒ ⓓ
15.	ⓐ ⓑ ⓒ ⓓ	48.	ⓐ ⓑ ⓒ ⓓ	82.	ⓐ ⓑ ⓒ ⓓ
16.	ⓐ ⓑ ⓒ ⓓ	49.	ⓐ ⓑ ⓒ ⓓ	83.	ⓐ ⓑ ⓒ ⓓ
17.	ⓐ ⓑ ⓒ ⓓ	50.	ⓐ ⓑ ⓒ ⓓ	84.	ⓐ ⓑ ⓒ ⓓ
18.	ⓐ ⓑ ⓒ ⓓ	51.	ⓐ ⓑ ⓒ ⓓ	85.	ⓐ ⓑ ⓒ ⓓ
19.	ⓐ ⓑ ⓒ ⓓ	52.	ⓐ ⓑ ⓒ ⓓ	86.	ⓐ ⓑ ⓒ ⓓ
20.	ⓐ ⓑ ⓒ ⓓ	53.	ⓐ ⓑ ⓒ ⓓ	87.	ⓐ ⓑ ⓒ ⓓ
21.	ⓐ ⓑ ⓒ ⓓ	54.	ⓐ ⓑ ⓒ ⓓ	88.	ⓐ ⓑ ⓒ ⓓ
22.	ⓐ ⓑ ⓒ ⓓ	55.	ⓐ ⓑ ⓒ ⓓ	89.	ⓐ ⓑ ⓒ ⓓ
23.	ⓐ ⓑ ⓒ ⓓ	56.	ⓐ ⓑ ⓒ ⓓ	90.	ⓐ ⓑ ⓒ ⓓ
24.	ⓐ ⓑ ⓒ ⓓ	57.	ⓐ ⓑ ⓒ ⓓ	91.	ⓐ ⓑ ⓒ ⓓ
25.	ⓐ ⓑ ⓒ ⓓ	58.	ⓐ ⓑ ⓒ ⓓ	92.	ⓐ ⓑ ⓒ ⓓ
26.	ⓐ ⓑ ⓒ ⓓ	59.	ⓐ ⓑ ⓒ ⓓ	93.	ⓐ ⓑ ⓒ ⓓ
27.	ⓐ ⓑ ⓒ ⓓ	60.	ⓐ ⓑ ⓒ ⓓ	94.	ⓐ ⓑ ⓒ ⓓ
28.	ⓐ ⓑ ⓒ ⓓ	61.	ⓐ ⓑ ⓒ ⓓ	95.	ⓐ ⓑ ⓒ ⓓ
29.	ⓐ ⓑ ⓒ ⓓ	62.	ⓐ ⓑ ⓒ ⓓ	96.	ⓐ ⓑ ⓒ ⓓ
30.	ⓐ ⓑ ⓒ ⓓ	63.	ⓐ ⓑ ⓒ ⓓ	97.	ⓐ ⓑ ⓒ ⓓ
31.	ⓐ ⓑ ⓒ ⓓ	64.	ⓐ ⓑ ⓒ ⓓ	98.	ⓐ ⓑ ⓒ ⓓ
32.	ⓐ ⓑ ⓒ ⓓ	65.	ⓐ ⓑ ⓒ ⓓ	99.	ⓐ ⓑ ⓒ ⓓ
33.	ⓐ ⓑ ⓒ ⓓ	66.	ⓐ ⓑ ⓒ ⓓ	100.	ⓐ ⓑ ⓒ ⓓ
		67.	ⓐ ⓑ ⓒ ⓓ		

POST-TEST

1. The supply department forgot to put the total on a recent invoice. If the cost of each of the three items was $12.56, $141.08, and $76.33, how much should the total bill be?
    a. $228.97
    b. $229.87
    c. $229.97
    d. $230.87

2. Which of the following represents 11% as a decimal?
    a. .011
    b. .11
    c. 1.1
    d. 11

3. At the start of the day, inventory reports showed that there were 37 drills in stock at the store. After a Father's Day sale, the receipts showed that nine drills were sold. How many drills were left in stock?
    a. 25
    b. 26
    c. 28
    d. 46

Use the figure below to answer questions 4–5.

Percentage of Couches by Upholstery

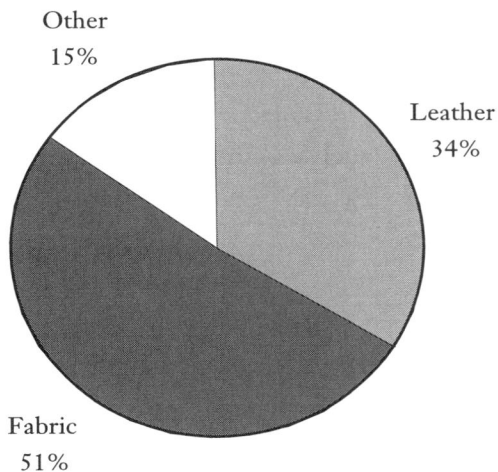

4. What is the percentage of couches that are not fabric?
   a. 15%
   b. 34%
   c. 49%
   d. 85%

5. How many couches were made in Fabric if there was a total of 1,300 couches produced?
   a. 663 couches
   b. 637 couches
   c. 442 couches
   d. 212 couches

6. A clerk accidentally scanned in an item twice, which cost $7.80. If the total bill with the mistake was $37.24, what should the correct bill be?
   a. $28.44
   b. $29.44
   c. $45.04
   d. $45.06

7. On a recent paycheck, Martha was paid $376.80 for a total of 18 hours of work. About how much does Martha earn per hour?
   a. $15.24
   b. $17.16
   c. $20.93
   d. $22.81

8. Each month $68.50 is taken out of Claire's paycheck for taxes, social security, and other government deductions. How much money per year is being sent to the government?
   a. $685.00
   b. $753.50
   c. $787.75
   d. $822.00

9. A plumber purchased 18 new fittings that cost $48 each. How much was the bill before tax?
   a. $754
   b. $834
   c. $864
   d. $944

10. In order to pay for a $16.54 bill, a customer hands the cashier a $50 bill. How much change is due back to the customer?

    a. $33.56

    b. $33.46

    c. $34.46

    d. $34.56

11. Anna is a purchaser in the ladies clothing department for a store. When she orders the inventory she must purchase $\frac{2}{5}$ of the stock for sizes 6–10 and $\frac{1}{3}$ of the stock for sizes 0–4. What fraction is left for sizes 12–16?

    a. $\frac{4}{15}$

    b. $\frac{5}{8}$

    c. $\frac{13}{15}$

    d. $\frac{3}{8}$

Use the figure below to answer questions 12–13.

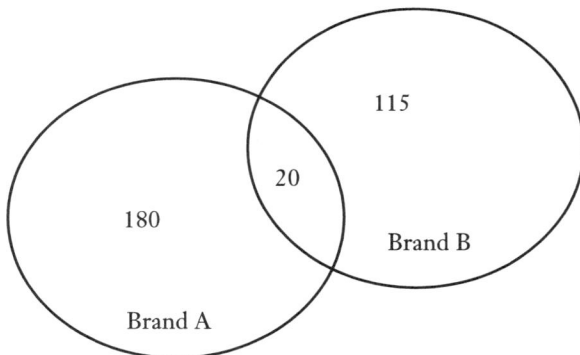

Above is an example of a Venn diagram. The numbers inside the circles represent how many consumers purchase each brand. The middle, overlapping section is for the customers who purchase both brands.

12. How many total products were purchased?

    a. 295

    b. 305

    c. 315

    d. 325

13. How many customers purchased Brand A?

   a. 115 Customers

   b. 135 Customers

   c. 180 Customers

   d. 200 Customers

14. Baseboards are sold in 16 foot sections. How many baseboards are necessary to complete a room with 322 linear feet of wall space?

   a. 18 boards

   b. 19 boards

   c. 20 boards

   d. 21 boards

15. How many CD players, costing $58 each, can be purchased with $854?

   a. 13

   b. 14

   c. 15

   d. 16

16. At a nursery there are baby palm trees on sale for $135 each. How much will it cost to purchase six baby palm trees?

   a. $675

   b. $810

   c. $840

   d. $945

17. Which drill bit is approximately 2.25″ long?

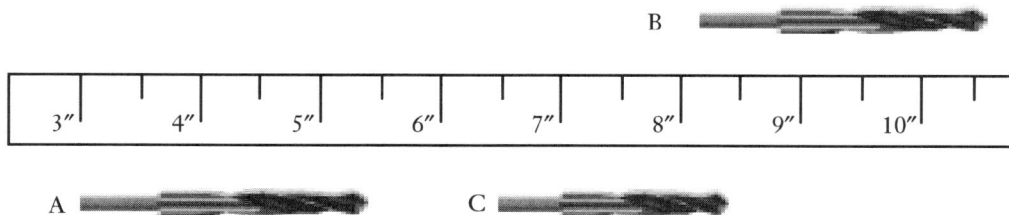

   a. A

   b. B

   c. C

   d. A and B

18. A manager is allotted $1,350 for materials on a project. The first purchase costs $579.50 and the second purchase costs $715.35. After this, $215 worth of materials was returned, and then a final purchase of $275.80 took place. How much money was left over at the end?

    a. $0

    b. $5.65

    c. $7.24

    d. The manager is over budget.

19. If a parking lot is $\frac{3}{4}$ full, what is this fraction represented as a percent?

    a. .75%

    b. 3.4%

    c. 25%

    d. 75%

20. You have 62.7 feet of rope. If 1 meter = 3.28 feet, approximately how many meters of rope do you have?

    a. 5.2 meters

    b. 19.1 meters

    c. 20.9 meters

    d. 33.5 meters

21. Decorative light switches are sold individually for $1.45 each, or sold in packages of 20 for $26.50. If 36 new switches are needed, what is the best combination to purchase?

    a. 2 packages of 20

    b. 1 package of 20 and 16 individual switches

    c. 36 individual switches

    d. 20 individual switches and 1 package of 20

22. You are trying to keep your employees from working overtime, which is anything over 40 hours in a week. If there is an employee who already has worked 9 hours on Monday, 6.5 hours on Tuesday, 8 hours on Thursday, and 10.5 hours on Friday, how many hours can you schedule that employee to work on Saturday?

    a. 5 hours

    b. 5.5 hours

    c. 6 hours

    d. 4.5 hours

23. There are 65 applicants for four job openings. Approximately how many people are applying per job?

    a. 15

    b. 16

    c. 17

    d. 18

24. Using the information from question 23, what is the probability of being hired for one of the jobs? Assume each person is equally qualified.
- a. .015
- b. .062
- c. .246
- d. .250

25. A cleaning service charges $75 for a job and pays the cleaner $42. How much profit does the cleaning service receive?
- a. $1.8
- b. $33
- c. $75
- d. $117

26. A realtor made 3.5% commission on the sale of a house. If the purchase price of the house was $224,500, how much money did the realtor make?
- a. $6,735.50
- b. $7,857.50
- c. $78,575.00
- d. $785,750.00

27. If the realtor in question 26 must give 25% of the earnings to the parent realty company, how much money does the realtor make on the sale?
- a. $1,964.38
- b. $3,015.89
- c. $4,645.26
- d. $5,893.13

28. Mark completed $\frac{1}{6}$ of his weekly hours of work on Saturday and then $\frac{2}{5}$ of his hours on Sunday. What fraction of his hours has he finished?
- a. $\frac{3}{11}$
- b. $\frac{17}{30}$
- c. $\frac{2}{30}$
- d. $\frac{17}{11}$

29. Using the situation in question 28, approximately what percentage of work does Mark have left in order to fulfill his weekly quota?
- a. 43%
- b. 57%
- c. 73%
- d. not enough information

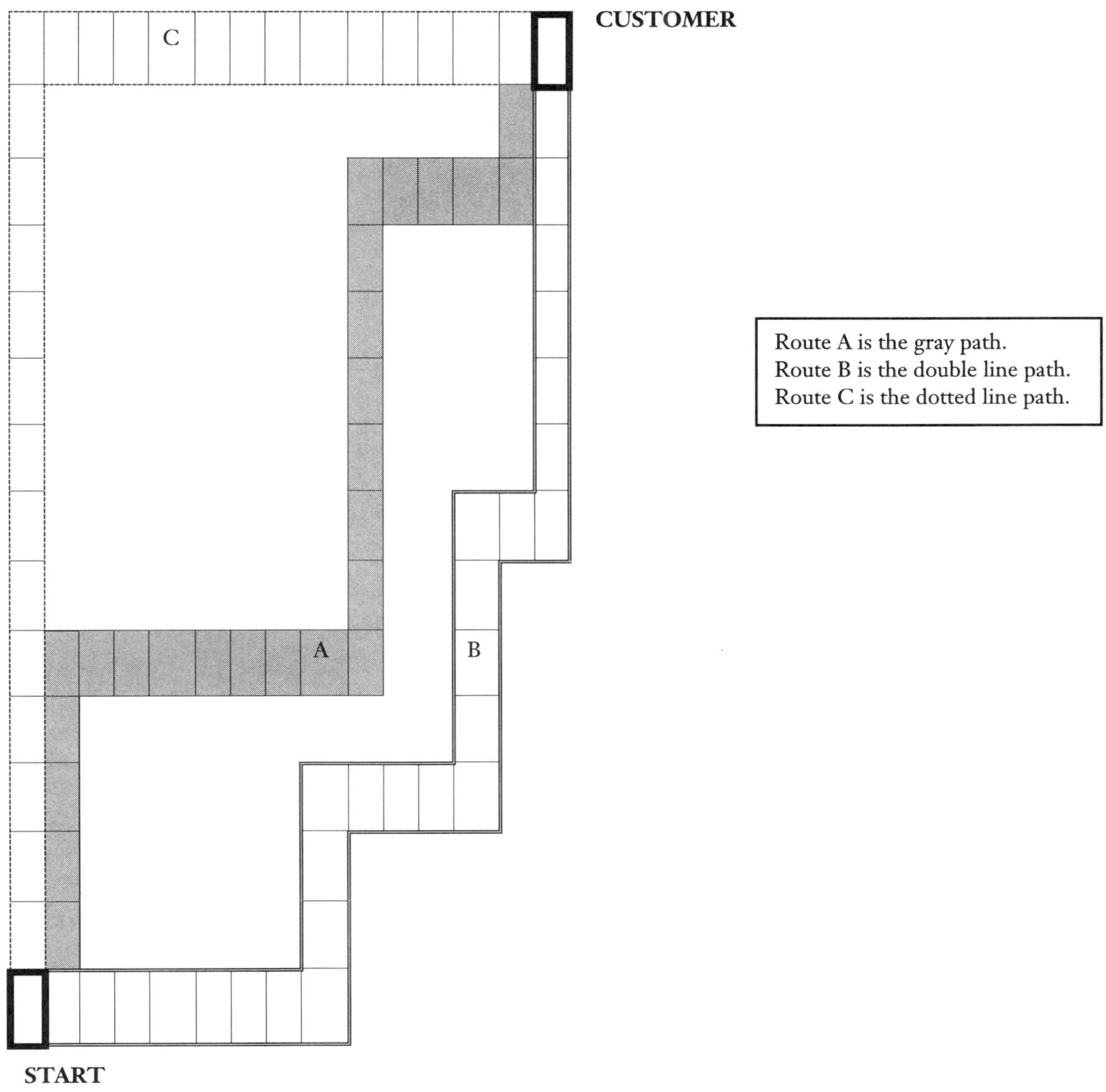
C
CUSTOMER
A
B
START
Route A is the gray path.
Route B is the double line path.
Route C is the dotted line path.

30. Using the above figure, which route to the customer has the shortest distance?

a. route A

b. route B

c. route C

d. route B and route C are the same.

31. A common workday is eight hours. What percentage of a day is represented by a workday?

a. 16.0%

b. 33.3%

c. 45.0%

d. 66.7%

32. The sales tax in a certain state is 6.2%. In that state, what is the total cost for a purchase that costs $30 before tax?

   a. $3.86
   b. $31.86
   c. $36.29
   d. $48.60

33. To mail a letter, postage costs 37 cents for one ounce and 23 cents for each additional ounce or fraction of an ounce. How much will it cost to send a letter that weighs 4.2 oz?

   a. $1.06
   b. $1.29
   c. $1.52
   d. $1.85

34. A delivery truck company damaged 12 barbeques out of a total shipment of 145 barbeques. What percentage of the barbeques was not damaged?

   a. 8.3%
   b. 13.3%
   c. 85.1%
   d. 91.7%

35. The following diagrams show an aerial view of two file cabinets. Which cabinet should be purchased to maximize floor space?

1.2 ft.   A   2.7 ft.

B   1.8 ft.   1.8 ft.

   a. file cabinet A
   b. file cabinet B
   c. either one
   d. not enough information

36. A tanker is $\frac{1}{10}$ empty. What percentage of the capacity is available?

   a. 10%
   b. 50%
   c. 90%
   d. 100%

37. For an electrical job, 22 new recessed can lights will be installed. The cans come in boxes of five and the baffles (rims) come in sets of eight. How many of each should you purchase in order to have the least amount of materials left over??
 a. 4 boxes of cans and 4 boxes of baffles
 b. 5 boxes of cans and 3 boxes of baffles
 c. 6 boxes of cans and 3 boxes of baffles
 d. 8 boxes of cans and 5 boxes of baffles

Use the figure below to answer questions 38–39.

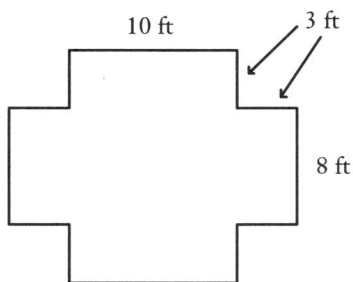

38. The courtyard in the figure above must first be lined before the bricklayers can complete it. What is the perimeter of the courtyard?
 a. 30 feet
 b. 52 feet
 c. 60 feet
 d. 64 feet

39. In order to lay the bricks to complete the courtyard, the bricklayer must find out the area. What is that area of the courtyard?
 a. 134 ft²
 b. 164 ft²
 c. 188 ft²
 d. 224 ft²

40. There are two lengths of pipe that will be fitted together. If one pipe is 6 ft 7 in and the other is 9 ft 11 in, how long will the new pipe be?
 a. 15 ft 6 in
 b. 15 ft 18 in
 c. 16 ft 6 in
 d. 17 ft 16 in

41. A housekeeping company charges a flat fee of $25/house. There is an additional cost of $4.50 per room, with the first room included in the flat fee. What is the total charge for cleaning two houses that have seven rooms each?

   a. $52

   b. $104

   c. $113

   d. $175

Use the table below to answer questions 42–43.

	DAY SHIFT	EVENING SHIFT	GRAVEYARD SHIFT
# of Employees	154	122	59
Tasks to Complete	385	164	155

The table above shows the typical work schedule for a power plant.

42. How many employees work at the power plant?

   a. 225

   b. 335

   c. 426

   d. 626

43. Which shift has to perform more tasks per person?

   a. day shift

   b. evening shift

   c. graveyard shift

   d. They are all the same.

44. Today, 175 new computers were supposed to be delivered. Unfortunately, 50 laptops were not delivered. Of those 50 late computers, 32 will be rush delivered the following day. What percentage of the 50 missing laptops will still be missing the following day?

   a. 10%

   b. 18%

   c. 36%

   d. 42%

45. A commercial box of nails weighs 34 lbs. If each nail weighs approximately 1 oz, how many nails are in the box? (1 lb = 16 oz. The weight of the box is negligible.)

   a. 34

   b. 544

   c. 2,176

   d. 4,352

46. Two-thirds of the remaining "last-year" model cars will be discounted at the end of the day. Each car will be discounted around $1,500 and there are still 291 last-year models available. What is the total dollar value discount that this price reduction will bring for all of the cars combined?

  a. $145,500
  b. $291,000
  c. $436,500
  d. $654,750

47. A car rental company is offering a 20% discount for customers during the week. If the weekday price quote was $35, what is the weekend price?

  a. $42.00
  b. $43.75
  c. $55.00
  d. $57.14

Use the figure below to answer questions 48–51.

Average Nightly Occupancy By Month

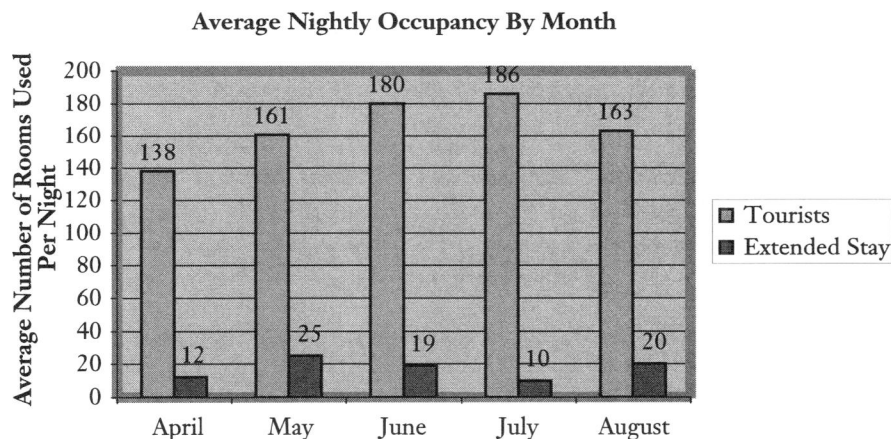

Tourists are defined as patrons who stay for one week or less. Extended stay guests are visitors who stay for more than one week.

48. Approximately how many nights were the rooms occupied in this hotel in May?

  a. 186 times
  b. 4,650 times
  c. 5,580 times
  d. 5,766 times

49. Which month has the highest average of occupancy?

   a. June
   b. July
   c. August
   d. not enough information

50. What percentage of rooms did Extended Stay guests use over the time period?

   a. 7.6%
   b. 9.4%
   c. 10.4%
   d. 73.3%

51. What is the percent increase of rooms used by tourists from April to August?

   a. 18.1%
   b. 119.6%
   c. 153.0%
   d. 330%

Use the chart below to answer questions 52–55. It shows what two Internet and Cable services companies are offering to their customers.

	COMPANY A	COMPANY B
Connection Fee	$12.95	$8.50
Cable	$26.75/month	$27.00/month
High-Speed Internet	$29.99/month	$30.50/month
Combo Package of Cable and High-Speed Internet (no connection fee when this service is selected)	$55.80/month	$54.00/month

52. For a year's subscription, which company offers a better deal for customers who only want cable?

   a. Company A
   b. Company B
   c. They offer the same deal.
   d. not enough information

53. For a year's subscription, which company offers a better deal for customers who only want high-speed Internet?
   a. Company A
   b. Company B
   c. They offer the same deal.
   d. not enough information

54. Which company offers a better deal on digital cable?
   a. Company A
   b. Company B
   c. They offer the same deal.
   d. not enough information

55. Which company offers the larger discount if a combo package is selected for a year?
   a. Company A
   b. Company B
   c. They provide the same discount.
   d. not enough information

56. A cylindrical tank is 10 ft tall with a diameter of 4 ft. What is the potential volume of the tank?

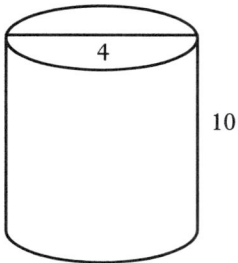

   a. 40 ft^3
   b. 54.5 ft^3
   c. 62.8 ft^3
   d. 125.6 ft^3

57. A chiropractor is offering a 45% discount for new customers on their first five visits. In order to cover costs, the normal $35 consultation fee will first be increased by 35%. If the consultation counts as their first visit, what is the consultation fee that the new customers will pay?
   a. $21.26
   b. $25.99
   c. $32.46
   d. $47.25

58. The base for a new fountain is in the shape of the figure below. How much area is needed to build the fountain?

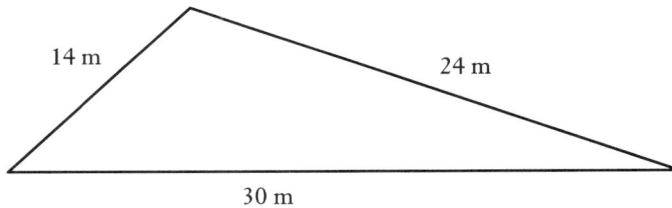

14 m 24 m 30 m

a. 168 m²
b. 210 m²
c. 360 m²
d. not enough information

59. If the fountain dimensions of the figure in question 58 are reduced by $\frac{1}{4}$, what is the new perimeter?
a. 17 m
b. 22 m
c. 51 m
d. 68 m

60. A canning plant received 1.5 tons of tuna that is to be packaged in 8 oz. cans. How many cans of tuna will be produced from this shipment?
a. 6,000
b. 7,000
c. 8,000
d. 9,000

61. Jeff is paid double rate for hours worked over his required 50 hours/week. What is his regular hourly wage if he made a total of $1,244.25 for a 64.5-hour workweek?
a. $15.75/hour
b. $16.45/hour
c. $16.85/hour
d. $19.29/hour

62. A small business owner has saved up a substantial amount of money to invest. He has $15,000 and would like to invest 40% of the money into a fund that returns 12% annually and the rest of the money in bonds that return 7% annually. How much money has he earned at the end of a year?
a. $1,350
b. $1,425
c. $1,500
d. $16,350

63. There are two gears that are connected together. One complete turn of the handle rotates the first gear exactly five times and the second gear rotates exactly two times. How many rotations of the handle will it take to get the first gear to rotate exactly 11 times?

    a. 6

    b. 3.6

    c. 2.5

    d. 2.2

64. Mike, Steve, and DJ are contractors with a total of 440 clients between them. Steve has $\frac{3}{8}$ the number of clients Mike has, and DJ has three times the number Steve has. How many clients does DJ have?

    a. 176

    b. 183

    c. 198

    d. not enough information

65. Of the contractors mentioned in question 64, who makes the most money?

    a. Mike

    b. Steve

    c. DJ

    d. not enough information

66. A staircase with 14 stairs is being constructed in a new house. Each stair has a ratio of 3 : 7 for the height and depth respectively. If each stair is 15 inches deep, what is the height of the entire staircase?

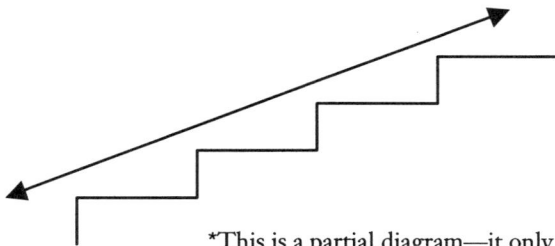

*This is a partial diagram—it only shows the first few stairs

    a. 7.5 ft

    b. 9.3 ft

    c. 17.5 ft

    d. 90 ft

67. Using the diagram in question 66, what is the diagonal length of the staircase (see arrow in the diagram)?

   a. 14.33 ft

   b. 18.72 ft

   c. 19.04 ft

   d. 25 ft

68. Scott, an electrician, charges an initial fee of $35 plus the cost of materials, plus his hourly rate of $65/hour for labor. Scott worked on Tuesday from 9:30 A.M. until 2:45 P.M. with a half an hour (unpaid) lunch break. On Wednesday, he worked from 9:00 A.M. until 4:30 P.M. with a 45-minute (unpaid) lunch break. How much will Scott charge for this job with materials costing $47.35?

   a. $794.85

   b. $782.50

   c. $829.85

   d. $911.10

69. A car rental company charges $45/week and an additional 5¢/mi. Which equation would calculate the weekly rental cost? (C = cost, m = miles)

   a. $C = 45 + 5m$

   b. $C = .05m + 45$

   c. $C = 45m + .05$

   d. $C = 5(m + 45)$

Use the following to answer questions 70–72.

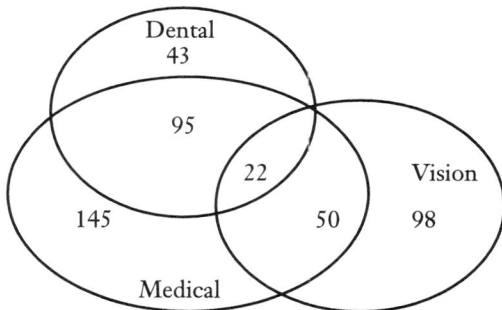

The Venn diagram above shows the different health benefit plans that the employees of Harrison Inc. are offered and the number of employees that chose each of the options.

70. How many employees work at Harrison Inc.?

   a. 286 employees

   b. 453 employees

   c. 512 employees

   d. not enough information

71. What percentage of the employees (on a health plan) have some kind of vision coverage?

    a. 32.7%

    b. 33.3%

    c. 37.5%

    d. 39.4%

72. If you randomly pick one employee who takes part in a health plan, what is the probability that they have some kind of dental coverage?

    a. .0625

    b. .35

    c. .42

    d. 34%

73. What is a reasonable average speed for a delivery truck that has 280 miles to cover in less than five hours?

    a. 52 mph

    b. 55 mph

    c. 59 mph

    d. 80 mph

Use the figure to answer questions 74–75.

Employee Distribution

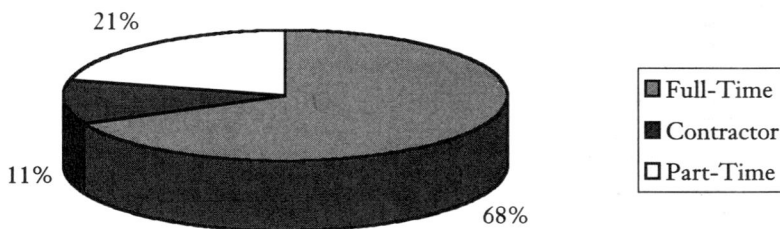

74. 68% of the full-time workers, 45% of the part-time workers, and 25% of the contractors are women. If there are 1,000 employees in the company, how many are women?

    a. 416

    b. 584

    c. 591

    d. 612

75. Due to budget restraints, the overall workforce has been reduced by 18%. Assuming there were originally 1,000 employees, and that the percentage of each type of employee remains the same after the layoffs, approximately how many full-time workers are there now?

a. 558

b. 612

c. 743

d. 900

76. A gas tanker started the day with 310 gallons in its tank, which was already $\frac{2}{5}$ empty. The first stop depleted the supply by $\frac{2}{9}$ and the second stop took $\frac{1}{3}$ of the remaining gas. At a fill-up, how much will the tanker take?

a. 264.1 gallons

b. 356 gallons

c. 426.6 gallons

d. 470.8 gallons

Use the following figure to answer questions 77–79.

Cost and Charge (in dollars) for Tune-Up, 2000-2003

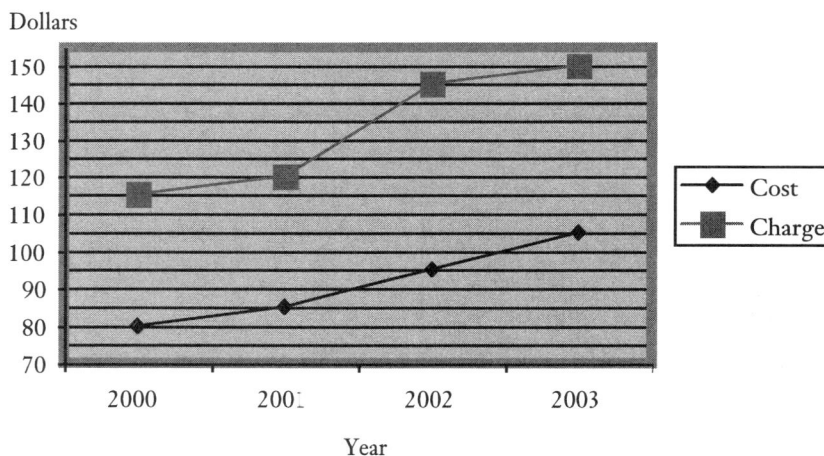

This graph represents the cost for an auto repair shop to provide an entire tune-up. The diamond line represents the actual cost to the shop to perform the service and the square line represents what they charged the customers.

77. Which year had the greatest margin of profit?

a. 2000

b. 2001

c. 2002

d. 2003

78. Which year provided the most money from tune-ups?
   a. 2001
   b. 2002
   c. 2003
   d. not enough information

79. If there was a profit of $9,590 for the tune-ups in 2001, how many tune-ups were performed in 2001?
   a. 80
   b. 113
   c. 198
   d. 274

80. How many pounds of cashews that cost $5.25/lb should be added to a 5 lb mixed nut mixture that currently costs $3.25/lb to end up with a mixture that costs $4.50/lb?
   a. 5 lbs
   b. 6.2 lbs
   c. 8.33 lbs
   d. 9.02 lbs

Use the figure to answer questions 81–82.

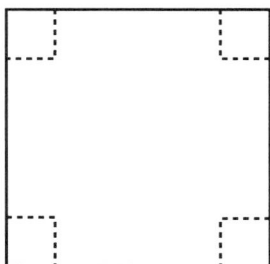

81. A square piece of sheet metal with a length of 16 inches must be cut and formed into an open box. If a 3-inch square is cut from each corner, the box can be formed. What is the volume that this new box can hold?
   a. 300 in^3
   b. 442 in^3
   c. 507 in^3
   d. 768 in^3

82. What is the difference in surface area of one side of the original piece of metal to the new cut piece?
   a. 36 in^2
   b. 48 in^2
   c. 54 in^2
   d. 220 in^2

83. Dana and Faith are pharmaceutical representatives and they work with 192 doctors. If Faith has $\frac{2}{3}$ more clients than Dana, how many doctors does Dana handle?

    a. 72

    b. 84

    c. 120

    d. 128

Use the figure to answer questions 84–85.

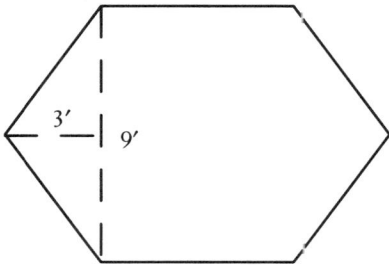

84. If the figure above represents a garden, what is the area?

    a. 67.5 ft²

    b. 81 ft²

    c. 108 ft²

    d. 114 ft²

85. What is the perimeter of the garden?

    a. 30 ft

    b. 33.6 ft

    c. 35.4 ft

    d. not enough information

86. A wall under construction needs a support to run from the top of the wall to the floor at an angle. The height of the wall is 9.4 ft and the length of the support is 16.5 ft. How far away from the wall will the support stand?

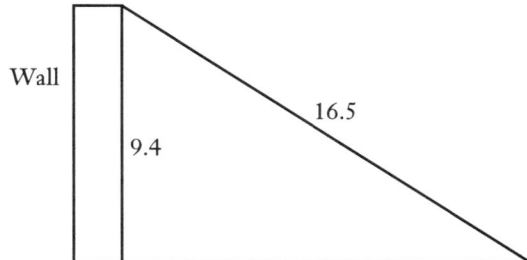

Wall

9.4

16.5

- **a.** 7.1 ft
- **b.** 9.5 ft
- **c.** 11.4 ft
- **d.** 13.6 ft

87. Mary purchased chairs for $38 and sold all but 20 of them for $64 each, making a profit of $358. How many chairs did she originally purchase?
- **a.** 43
- **b.** 63
- **c.** 88
- **d.** not enough information

88. Matt purchased 100 faucets from the distributor at $29 each. From the store data, he is guaranteed to sell at least 80 faucets in the next three months. How much should he price the faucets for if he wants to make at least $3,870 profit from selling them?
- **a.** $78.45
- **b.** $81.95
- **c.** $82.15
- **d.** $84.65

Use the following figure to answer questions 89–91.

Average Car Sales Per Month (2001-2003) for Becky, Curtis, and Austin

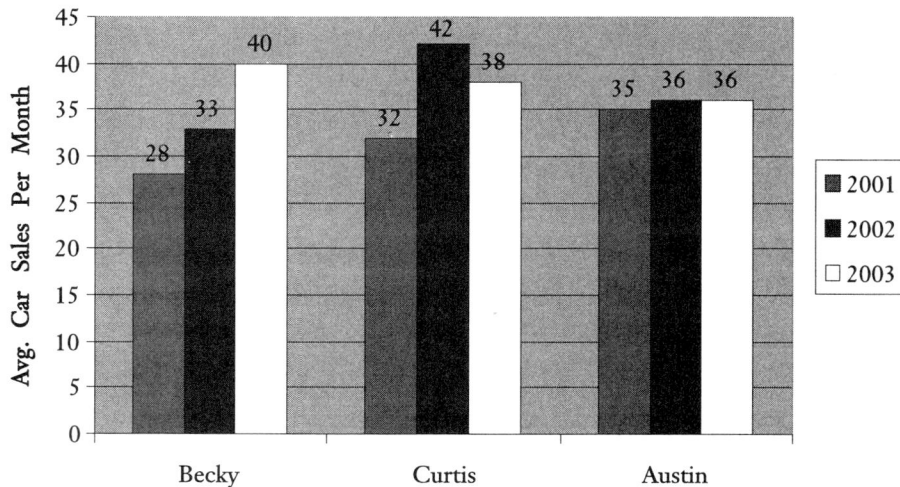

89. The graph shows the average monthly car sales from the past three years for three of the salespersons at a car dealership. Approximately how many cars were sold in 2002 by these individuals?
   a. 1,298
   b. 1,332
   c. 1,412
   d. 1,544

90. Comparing the yearly sales, which of the following had the largest increase from the previous year?
   a. 2001
   b. 2002
   c. 2003
   d. They all showed decreasing sales.

91. Becky sold 25 cars per month in 2000. Looking at Becky's average increase, what are her sales likely to be in 2004 if she continues her trend?
   a. 39 per month
   b. 42 per month
   c. 43 per month
   d. 49 per month

92. A computer will be marked up 15% before the 25% off sale. How much will the computer cost if the original price was $1,850 and there is a sales tax of 6.5%?

   a. $1,543.89

   b. $1,605.41

   c. $1,699.35

   d. $1,785.25

93. The ratio of 6 : 3 : 1 represents the inventory of 6-foot, 8-foot, and 12-foot pieces of wood, respectively. If there are a total of 1,400 pieces of wood, how many 8-foot pieces are there?

   a. 140

   b. 280

   c. 420

   d. 840

94. How much fabric is required to cover the box below if the length is 2.5 ft, the width is 8 in, and the height is 3 in?

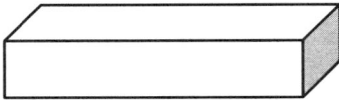

   a. 354 in^2

   b. 452 in^2

   c. 708 in^2

   d. 2,160 in^2

95. Lacey can clean a hotel room in 14 minutes alone and Maria can clean a hotel room in 19 minutes. If Lacey and Maria work together, how many complete rooms can they clean in one hour?

   a. 6 rooms

   b. 7 rooms

   c. 8 rooms

   d. 9 rooms

Use the following figure to answer questions 96 and 97.

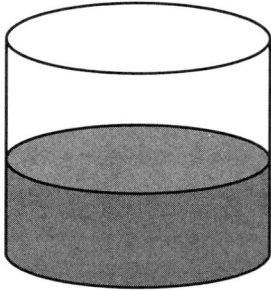

96. The cylindrical container is $\frac{2}{5}$ full of grain and has a capacity to hold an additional 217.5 m³. How much grain is already in the container?
   a. 87 m³
   b. 145 m³
   c. 186.2 m³
   d. 362.5 m³

97. If the diameter of the container is 8.5 cm, what is the approximate height of the entire container?
   a. 1.6 m
   b. 3.8 m
   c. 6.3 m
   d. 7.5 m

98. A certain telephone customer has two long distance plans from which to choose. Option 1 costs $7/month for 200 minutes and 5¢ for each minute over the allotted 200 minutes. Option 2 costs 7¢/min with no monthly charge. Which is a better deal for a customer who only talks 140–155 minutes per month?
   a. Option 1
   b. Option 2
   c. They would both cost the same.
   d. not enough information

99. The following is a drawing of a house that is to be constructed. The scale of the drawing is 2 in : 5 ft. If the house in the drawing is 10.2″ tall, how tall will the actual house be?

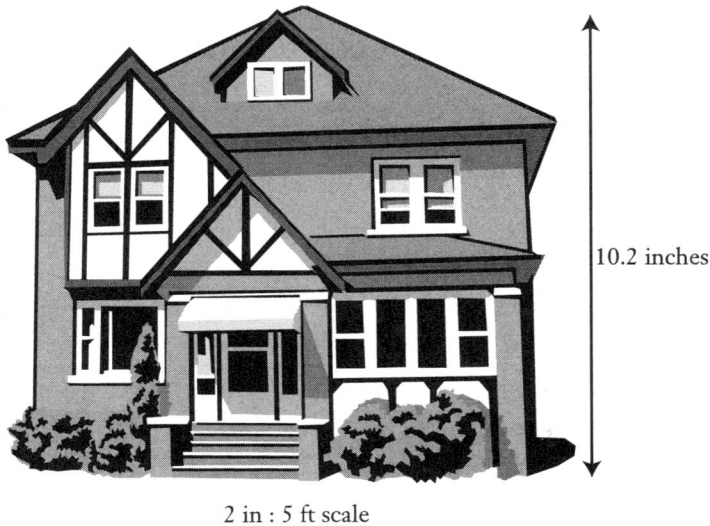

10.2 inches

2 in : 5 ft scale

    a. 22.9 ft
    b. 25.5 ft
    c. 26.8 ft
    d. 29.4 ft

100. There is a rectangular pool that is 35 m × 80 m. A 2.5 m wide pebble walkway is going to be made around the pool. What is the area of the pebble walkway?

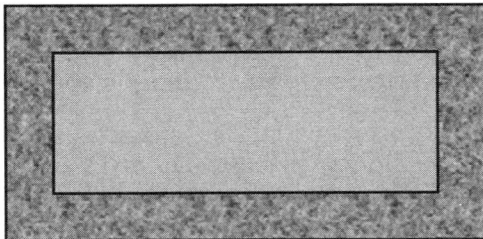

    a. 600 m^2
    b. 900 m^2
    c. 2,800 m^2
    d. 3,400 m^2

ANSWERS

1. **c.** To find the total bill, add up the cost of the items for a total of $229.97 (12.56 + 141.08 + 76.33 = 229.97).

2. **b.** To switch from percentages to decimals, divide the percentage by 100 ($\frac{11}{100}$ = .11). You can also move the decimal point two places to the left.

   0.1 1

3. **c.** To find out how many are left, subtract the original amount by the amount that was sold (37 − 9 = 28).

4. **c.** You can add up the other categories to find the percentage of couches that are not Fabric (15 + 34 = 49%). Or you can subtract the percentage that is represented by Fabric from 100% (100 − 51 = 49).

5. **a.** The object is to find 51% of the 1,300 couches. To find 51% of the 1,300 couches, simply multiply, remembering to put 51 over 100 because it represents a percentage (1,300 × $\frac{51}{100}$ = 663).

6. **b.** The customer has been overcharged by $7.80, so subtract this mistake from the total bill (37.24 − 7.80 = 29.44).

7. **c.** To find out how much Martha earns in an hour, divide the total money by the hours to find out "money per hour" ($\frac{376.80}{18}$ = 20.93).

8. **d.** There are 12 months in a year and $68.50 is held each month for the government. To find the total for the year, multiply the cost each month by 12 to represent an entire year (12 × 68.50 = $822).

9. **c.** When multiple items are purchased for the same price, simply use multiplication (48 × 18 = 864).

10. **b.** Calculating change requires subtraction. Take the money the customer gives and subtract away the total bill (50 − 16.54 = 33.46).

11. **a.** First you must figure out what fraction is already used from the first two categories of sizes. In order to add fractions, find a common denominator, which is 15 for this problem ($\frac{2}{5}$ + $\frac{1}{3}$ = $\frac{2(3)}{15}$ + $\frac{1(5)}{15}$ = $\frac{6+5}{15}$ = $\frac{11}{15}$). To find what is left, over subtract this fraction from 1, which represents the entire purchase (1 − $\frac{11}{15}$ = $\frac{15}{15}$ − $\frac{11}{15}$ = $\frac{4}{15}$).

12. **c.** Add up the total from each section, Brand A, Brand B, and both brands (180 + 20 + 115 = 315).

13. **d.** To find the total for Brand A, add up the customers who only purchased Brand A and the customers who purchased both brands (180 + 20 = 200).

14. **d.** Take the total distance and divide it by the length of the board ($\frac{322}{16}$ = 20.125) and then you will need to round up to 21. Twenty boards is not enough because it only covers (20 × 16 = 320) 320 linear feet.

15. b. Take the total money and divide it by the cost to find out how many can be purchased ($\frac{854}{58}$ = 14.72). For this problem you must round down to 14 CD players. If you try to purchase 15 CD players the total bill will be too high ($15 \times 58 = 870$).

16. b. For multiple items, take the cost per item and multiply it by the number of items purchased ($135 \times 6 = 810$).

17. d. Drill bit A is around 2.25″ long—you can tell this by counting from the 3″ spot (1″ to the 4″ marker, 1″ more to the 5″ marker). Then it is about halfway to the $5\frac{1}{2}$″ marker, which is an additional .25″ for a total of 2.25″; $1 + 1 + .25 = 2.25$. Drill bit C is only 2″ long. You can count using the ruler or subtract the ending mark from the starting mark ($8.5 - 6.5 = 2$). Drill bit B is also 2.25″ and you can find that by either adding up the distances or subtracting the ending place (10.5) from the beginning place (8.25) to find that it is also around 2.25″ ($10.5 - 8.25 = 2.25$).

18. d. Find out how much money the manager has spent and compare it to his budget ($579.50 + 715.35 - 215 + 275.80 = 1,355.65$). The manager only had \$1,350, so he is over budget by \$5.65 ($1,355.65 - 1,350 = 5.65$).

19. d. To convert from a fraction to a percent, simply multiply the fraction by 100 and simplify ($\frac{3}{4} \times 100 = \frac{300}{4} = 75\%$).

20. b. When converting from feet to meters (smaller unit → larger unit), divide the total feet by 3.28 ($\frac{62.7}{3.28} = 19.1$ meters).

21. b. One package and 16 individual switches costs \$49.70 ($26.50 + 16(1.45) = 26.50 + 23.30 = 49.70$). The first option (2 packages) costs \$53.00 ($2 \times 26.50 = 53$). Purchasing 36 individual switches is not economical and costs 52.20 ($36 \times 1.45 = 52.50$). The last option has too many switches and also costs more ($20 \times 1.45 + 26.50 = 29 + 26.50 = 55.50$).

22. c. Take the total time allowed (40 hours) and subtract the time already used ($9 + 6.5 + 8 + 10.5 = 34$) to find the leftover time ($40 - 34 = 6$).

23. b. To find out how many people are applying for each job, divide the total number of applicants by the total number of positions available ($\frac{65}{4} = 16.25$). The answer is closer to 16 than 17 people.

24. b. To find the probability of getting one of the jobs, take the total number of jobs available and divide it by the total applicants ($\frac{4}{65} = .062$), which is about a 6.2% chance of getting one of the jobs.

25. b. To find the profit, subtract the amount the company pays its worker (\$42) from the total money charged (\$75) to result in \$33 ($75 - 42 = 33$).

26. b. Find out what 3.5% of \$224,500 is by multiplying them together, remembering to divide the percentage by 100 ($224,500 \times \frac{3.5}{100} = 7,857.50$).

27. d. The realtor earned \$7,857.50 and now must give up 25%. If the realtor is giving up 25% then 75% of the commission is what she gets to keep. Calculate 75% of 7,857.50 by multiplying them together and dividing by 100 because of the percentage ($7,857.50 \times \frac{75}{100} = 5,893.13$).

28. b. Take the fractions that Mark has already worked and add them together by finding a common denominator ($\frac{1}{6} + \frac{2}{5} = \frac{1(5)}{30} + \frac{2(6)}{30} = \frac{5 + 12}{30} = \frac{17}{30}$).

29. a. Mark finished $\frac{17}{30}$, which is also 56.7% (divide 17 by 30 and multiply by 100: $\frac{17}{30} \times 100 = 56.7$). In order to find what is left over to do, subtract what he has finished from 100% (100 – 56.7 = 43.3%).

30. a. Count the squares that each path takes and choose the one with the smaller number. Route A—25, Route B—27, Route C—27.

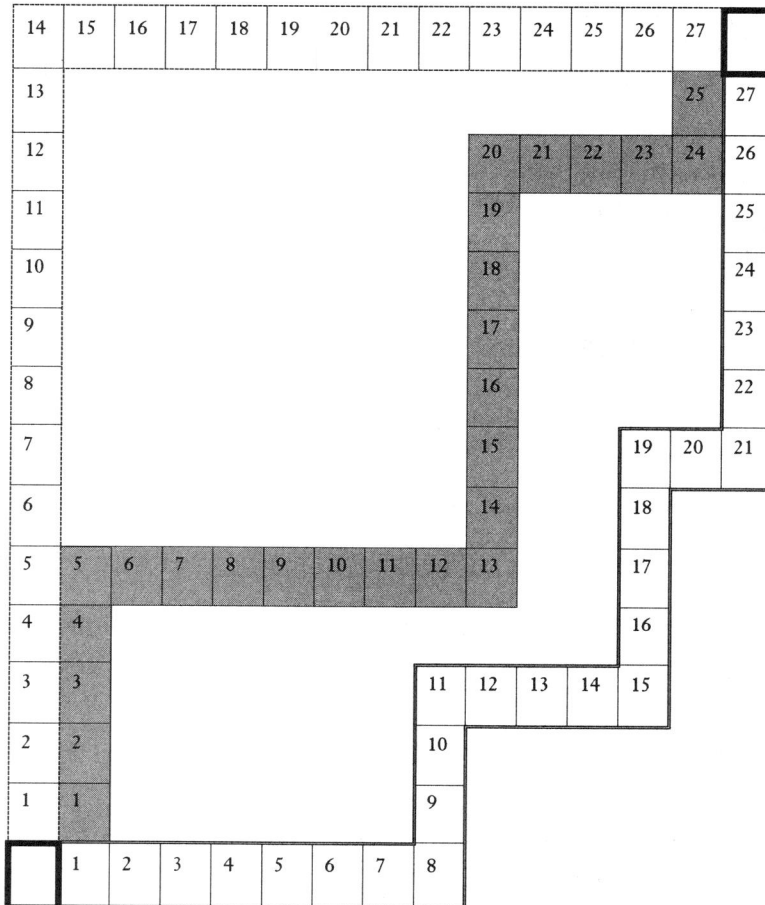

14	15	16	17	18	19	20	21	22	23	24	25	26	27	
13												25	27	
12									20	21	22	23	24	26
11									19					25
10									18					24
9									17					23
8									16					22
7									15			19	20	21
6									14			18		
5	5	6	7	8	9	10	11	12	13			17		
4	4											16		
3	3							11	12	13	14	15		
2	2							10						
1	1							9						
	1	2	3	4	5	6	7	8						

> Route A is the gray path.
> Route B is the double line path.
> Route C is the dotted line path.

31. b. One day has 24 hours and you must calculate what percentage of a day is taken up by working. Divide a workday (8) by an entire day (24) and then turn it into a percentage by multiplying by 100 ($\frac{8}{24} \times 100 = 33.3\%$).

32. b. To calculate the tax, multiply the cost (30) by the decimal value of the tax (.062) and add that to the original item ($30 \times .062 = 1.86 + 30 = \$31.86$). You are not just trying to find out the tax, but how much it costs altogether.

33. b. The letter weighs 4.2 oz. The first ounce is covered in the initial charge of .37 and the rest of the 3.2 ounces must be calculated with the additional charge. For postage they round up to the next ounce so the 3.2 ounces rounds up to 4 ounces. An additional 4 ounces will cost $.92 ($4 \times 23 = 92¢ = \$.92$). The total cost is .37 + .92 = $1.29.

34. d. The number of undamaged barbeques is 133 (145 – 12 = 133). To find the percentage available, divide the number of available barbeques by the total number that were sent and multiply by 100 to get a percentage ($\frac{133}{145} \times 100 = 91.7\%$).

35. **c.** The area of A is found by multiplying length (2.7) by width (1.2) for a total of 3.24 ft². File cabinet B's area is calculated the same way by multiplying the length (1.8) by the width (1.8) for a total of 3.24 ft². Each file cabinet uses the same area even though they have slightly different shapes.

36. **c.** The fraction $\frac{1}{10}$ is the same as 10% ($\frac{1}{10} \times 100 = 10\%$). 10% is the amount that is used; so 90% must be available (100% − 10% = 90%).

37. **b.** Find the smallest number of boxes for each part. 22 cans are needed and are sold in sets of 5, so 5 boxes are necessary (5×5 provides 25 cans). 22 baffles are also needed and they are sold in sets of 8, so only 3 boxes are necessary (3×8 provides 24 baffles). It is okay to have different amounts of each, as long as there are enough supplies for the job.

38. **c.** Perimeter is calculated by adding up all of the sides. For this figure the total is 60 (10 + 3 + 3 + 8 + 3 + 3 + 10 + 3 + 3 + 8 + 3 + 3 = 60).

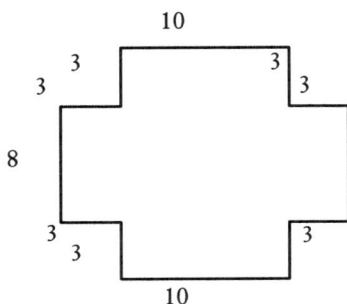

```
              10
        3           3
      3               3

    8

        3           3
          3       3
              10
```

39. **c.** The area of the courtyard can be calculated by breaking up the shape into smaller rectangles and calculating the area of each one, then adding them together for the total. Calculate the area by multiplying the length × width in each case (30 + 24 + 80 + 24 + 30 = 188).

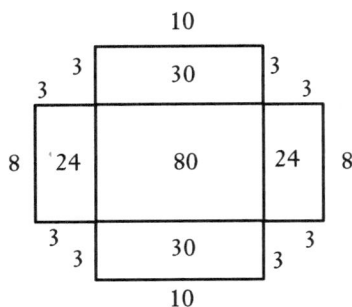

```
              10
        3           3
      3    30         3

    8  24    80   24  8

        3           3
          3  30   3
              10
```

40. **c.** Add the two pipes together and then simplify the inches (6 ft 7 in + 9 ft 11 in = 15 ft 18 in). It is not proper to leave the 18 in as it is; 18 in is the same as 1 ft 6 in (there are 12 inches in a foot, so subtract 12 from 18 to find that there are 1 ft 6 in in 18 in). Add 15 ft + 1 ft 6 in to get 16 ft 6 in.

41. **b.** The final calculation is for two houses, so it may be easier to find out how much it costs to clean one house and then double it at the end. One room in the house costs $25 so the additional six rooms will cost $27 ($6 \times 4.50 = 27$) so the total for one house is $52 (25 + 27 = 52). We are trying to find the cost for two houses, so double the cost for one house to find the total cost is $104 ($52 \times 2 = 104$).

42. **b.** To find the total number of employees, add up the total number in the row with the number of employees (154 + 122 + 59 = 335).

43. **c.** To find the tasks per person, divide the tasks by the employees. The day shift has 2.5 tasks per person ($\frac{385}{154}$ = 2.5), the night shift has 1.34 ($\frac{164}{122}$ = 1.34) and the graveyard shift has 2.63 ($\frac{155}{59}$ = 2.63). The graveyard shift has a higher number so more tasks are performed per person.

44. **c.** 32 computers are being delivered the next day, so 18 are still unaccounted for (50 – 32 = 18). Calculate the percentage out of 50 that these 18 computers represent ($\frac{18}{50}$ × 100 = 36%).

45. **b.** Use the conversion to calculate; if 1 lb = 16 oz, then multiply 34 by 16 to find the total number of oz (34 × 16 = 544). If each nail weighs 1 oz, there are 544 nails in a 34 lb box.

46. **b.** First find the number of cars that will have a discounted price. $\frac{2}{3}$ of the 291 cars is 194 ($\frac{2}{3}$ × 291 = 194). Each car will be discounted $1,500 for a total discount of $291,000 (194 × 1,500 = $291,000).

47. **b.** If a discount of 20% is offered, then the $35 quote represents 80% of the original price (100% – 20% = 80%). Translate the following information into an equation to solve: 80% of the original price equals $35. Use p as the original price and solve the equation; $\frac{80}{100}$ × p = 35 → .80p = 35 → $\frac{.80p}{.80}$ = $\frac{35}{.80}$ → p = 43.75.

48. **d.** Rooms are used for both tourists and extended stay guests. On an average day in May there are 186 rooms used. There are 31 days in May so the rooms are used 5,766 times (31 × 186 = 5766).

49. **a.** To find the average occupancy, add up the average tourist and extended stay rooms used. In April—150, May—186, June—199, July—193, and August—183. June has the highest average.

50. **b.** To find the percentage of rooms used by extended stay guests, find out the total average over the period of time shown (12 + 25 + 19 + 10 + 20 = 86). Also, find the total number of rooms used for regular tourists (138 + 161 +180 + 183 + 163 = 825) in order to find the total number of rooms used (86 + 825 = 911). Take the rooms used by extended stay guests and divide it by the total number of rooms used and multiply it by 100 to get the percentage ($\frac{86}{911}$ × 100 = 9.4%).

51. **a.** The tourists in April (138) can represent 100% of the customers. Calculate the percentage that the August customers represent (163) by dividing it by the guests in April and multiplying by 100 to get a percentage ($\frac{163}{138}$ × 100 = 118.1%). The percent increase is only 18.1% because the percentage went from 100% to 118.1%.

52. **b.** Cable costs for Company A for a year include the connection fee and 12 months of cable for a total of $333.95 (12.95 + 12(26.75) = 12.95 + 321 = $333.95). Company B costs 332.50 (8.50 + 12(27) = 8.50 + 324 = $332.50) and offers the better deal.

53. **a.** Internet costs for Company A for a year include the connection fee and 12 months of service for a total of $372.83 (12.95 + 12(29.99) = 12.95 + 359.88 = $372.83). Company B costs $374.50 (8.50 + 12(30.50) = 8.50 + 366 = 374.50). Company A offers a better deal by a few dollars.

54. **d.** There is no information about Digital cable so there is no way to compare what the companies offer.

55. **b.** Without a discount, it would cost $693.83 using Company A for an entire year (12.95 + 12(26.75) + 12(29.99) = 12.95 + 321 + 359.88 = 693.83). With a discount it would only cost $669.60

$(12 \times 55.80 = 669.60)$, for a savings of $24.23 ($693.83 − 669.60 = 24.23$). Without a discount, it would cost $698.50 using Company B for an entire year $(8.50 + 12(27) + 12(30.50) = 8.50 + 324 + 366 = 698.50)$. With a discount it would only cost $648 $(12 \times 54 = 648)$, for a savings of $50.50 $(698.50 − 648 = 50.50)$. Company B offers a larger discount from their normal prices compared to their discount/package offer.

56. **d.** Volume of a cylinder is calculated by finding the area of the base and multiplying it by the height of the cylinder. The area of the base is a circle and is calculated by $\pi r^2 = A$. The radius is half of the diameter, use 3.14 as an approximation for π $(3.14 \times (2)^2 = 3.14 \times 4 = 12.56)$. The entire volume is 125.6 ft^3 $(12.56 \times 10 = 125.6)$.

57. **b.** To figure out what the "discounted" consultation fee will be, first calculate the increase of the old consultation fee. To find a 35% increase on the $35, multiply the cost by 135% $(35 \times \frac{135}{100} = 47.25)$. The 45% discount will be offered on the adjusted price of $47.25. If there is a discount of 45%, the customers will be paying 55% of the cost $(100\% − 45\% = 55\%)$. 55% of $47.25 is $25.99 $(\frac{55}{100} \times 47.25 = \$25.99)$.

58. **d.** There is not enough information because the height of the triangle is unknown. To calculate the area of a triangle, use the formula $A = \frac{1}{2}bh$. The drawing gives all of the lengths, but no height is provided. The drawing gives all of the lengths of the sides of the triangle, but not the height. Below is an example of where a possible height might be located, but no value for this or any height is obtainable.

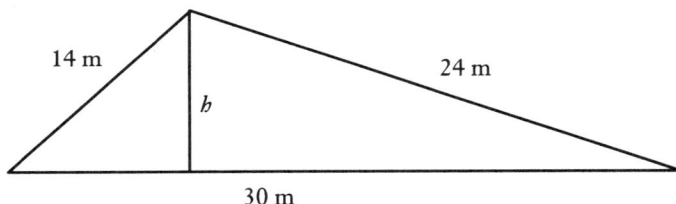

59. **c.** The current perimeter of the triangle is 68 m $(14 + 24 + 30 = 68)$. The perimeter must be reduced by $\frac{1}{4}$ for a reduction of 17 m $(\frac{1}{4} \times 68 = 17)$. To find the new perimeter, subtract the reduction from the larger perimeter for a new total of 51 m $(68 − 17 = 51$ m$)$.

60. **a.** 1.5 tons of tuna is the same as 3,000 lbs $(1.5 \times 2,000 = 3,000)$, and 3,000 lbs is the same as 48,000 ounces $(3000 \times 16 = 48,000)$. If each can holds 8 oz, then 6,000 cans will be produced $(\frac{48,000}{8} = 6,000)$.

61. **a.** Let Jeff's regular rate be x and his overtime rate be $2x$. He worked a total of 64.5 hours, so he worked his regular 50 hours plus 14.5 hours of overtime $(64.5 − 50 = 14.5)$. The money he earned from the regular week is time × rate, which is represented by $50x$ $(50 \times x = 50x)$. The money he earned from overtime work is also time × rate, which is $29x$ $(14.5 \times 2x = 29x)$. His regular salary plus overtime gives the total and can be written as $50x + 29x = 1,244.25$. Solve for x:

$$79x = 1,244.25$$
$$\frac{79x}{79} = \frac{1,244.25}{79}$$
$$x = 15.75$$

So Jeff makes $15.75/hour.

62. a. There is $15,000 to invest, 40% of which is invested at 12%; 40% of $15,000 is $6,000 (.40 × 15,000 = 6,000) so the remainder of the money is $9,000 (15,000 − 6,000 = 9,000). To find the interest made, multiply the money invested by the rate that is given. For the $6,000 the rate is 12% so the money returned is $720 (.12 × 6,000 = 720) and for the $9,000 the rate is 7% so the money returned is $630 (.07 × 9,000 = 630). To find the total money earned, simply add together the interest earned from each (720 + 630 = 1,350).

63. d. There is a ratio of handle to gear of 1 : 5. The gear needs to turn 11 times, so divide the total number of rotations by the number each handle turn makes ($\frac{11}{5}$ = 2.2) to find out how many rotations are needed.

64. c. This problem is simplified when an equation is used. Use x to represent the number of clients Mike has, $\frac{3}{8}x$ to represent Steve's clients, and $3(\frac{3}{8}x)$ to represent the clients for DJ. The total number of clients is 440, so add them up and solve for x.

$$x + \frac{3}{8}x + \frac{9}{8}x = 440$$
$$\frac{8}{8}x + \frac{3}{8}x + \frac{9}{8}x = 440$$
$$\frac{8}{20} \times \frac{20}{8}x = 440 \times \frac{8}{20}$$
$$x = 176$$

The question asks for how many clients DJ has, so substitute in 176 for x in $3(\frac{3}{8}x)$ to find that DJ has 198 clients.

65. d. There is no information about how much each makes. Just because Steve has the fewest clients does not mean he has the least amount of money coming into his firm. Even though DJ has the most clients, do not assume that he makes the most money.

66. a. There are 14 stairs with a ratio of 3 : 7. The total depth is 210 in (14 × 15 = 210) and using a simple proportion will help find the height.

$$\frac{3}{7} = \frac{x}{210}$$
$$7x = 3(210)$$
$$\frac{7x}{7} = \frac{630}{7}$$
$$x = 90$$

90 inches is the same as 7.5 ft ($\frac{90}{12}$ = 7.5).

67. c. The length of the staircase can be found by using the Pythagorean theorem. In the previous problem we found the depth and the height as shown below (this method of solution will use inches first and then convert to feet in the last step to match the answer choices).

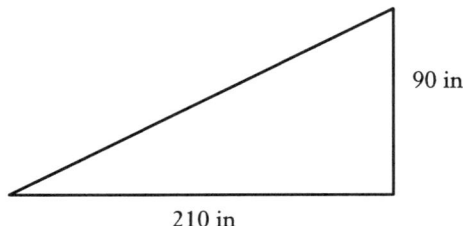

To find the hypotenuse, solve the following equation:

$$(210)^2 + (90)^2 = c^2$$
$$44,100 + 8,100 = c^2$$
$$52,200 = c^2$$
$$\sqrt{52,200} = \sqrt{c^2}$$
$$228.5 \approx c$$

The length of the staircase is about 228.5 in, which is the same as 19.04 feet ($\frac{228.5}{12} = 19.04$).

68. c. Scott worked 4.75 hours on Tuesday (5.25 total hours – .5 lunch = 4.75) and he worked 6.75 hours on Wednesday (7.5 total hours – .75 lunch = 6.75) for a total of 11.5 hours (4.75 + 6.75 = 11.5). He charges $65/hour, so for labor alone he charged $747.50 (11.5 × 65 = 747.50) for this job. Add the initial fee of $35 and the materials ($47.35) to get the total (747.50 + 35 + 47.35 = 829.85).

69. b. To find the cost of renting a car, start with the weekly rate ($45) and add that to the additional cost of the miles traveled. There is a charge of 5¢/mi, and since we have *m* miles in this problem, .05*m* represents the additional charge. Make sure to put 5¢ into dollar form so that all the information is in the same unit (the weekly charge is already in terms of dollars so it is easier to transfer everything into dollars). The total charge is found by adding the parts together 45 + .05*m*.

70. d. There is no information about employees who are not involved with a health plan, so there is no way to know from this diagram the total number of employees.

71. c. There are a total of 453 employees who have some kind of healthcare (43 + 95 + 145 + 22 + 50 + 98 = 453) and there are 170 employees enrolled in vision coverage (98 + 50 + 22 = 170). To find the percentage, divide the vision plan employees by the total in health care and multiply by 100 ($\frac{170}{453} \times 100 = 37.5\%$).

72. b. To find the probability of finding an employee with dental coverage, find the number of employees who have some form of dental coverage (43 + 95 + 22 = 160). The probability is simply taking the number of potential employees and dividing by the total number of employees enrolled in a health care plan ($\frac{160}{453} = .35$).

73. c. To find an average rate for distance, take the total distance and divide it by the total time. The classic equation is distance = rate × time, so rate = $\frac{\text{distance}}{\text{time}}$. The average rate for the truck is 56 mph ($\frac{280}{5} = 56$) if the parameters are followed exactly, but anything faster will also work.

The answer **c** is the most reasonable because it is not as safe for delivery trucks to drive 80 mph and the truck will make it in time driving 59 mph.

74. b. First find the number of employees in each category by multiplying the percentage by the total number of employees (Full-time: $.68 \times 100 = 680$, Part-time: $.21 \times 1,000 = 210$, Contractor: $.11 \times 1,000 = 110$). The percentage of women full-time employees is 68% so find 68% of the number of employees who are full-time ($.68 \times 680 = 462.4$) and the same with the other categories (Part-time: $.45 \times 210 = 94.5$, Contractor: $.25 \times 110 = 27.5$). There are a total of about 584 women ($462.4 + 94.5 + 27.5 = 584.4$).

75. a. The workforce is reduced by 18%, which means that 82% of the original 1,000 employees remains. The new number of employees is 820 ($.82 \times 1,000 = 820$). Since 68% of the remaining 820 employees are full-time, approximately 558 full-time employees still remain ($.68 \times 820 = 557.6$).

76. b. The tank was $\frac{2}{5}$ empty, which means it was $\frac{3}{5}$ full ($1 - \frac{2}{5} = \frac{5}{5} - \frac{2}{5} = \frac{3}{5}$). $\frac{3}{5}$ of the total capacity is equal to 310 gallons so the capacity of the tanker is 516.7 gallons ($\frac{3}{5}x = 310$, $x = 310 \times \frac{5}{3}$, $x = 516.7$). The first stop took $\frac{2}{9}$ of the original 310 which means 241.1 gallons remain ($\frac{2}{9} \times 310 = 68.9$, $310 - 68.9 = 241.1$). The next stop took $\frac{1}{3}$ of the remaining 241.1 gallons, resulting in 160.7 gallons left over ($\frac{1}{3} \times 241.1 = 80.4$, $241.1 - 80.4 = 160.7$). For the fill-up at the end, the truck will need 356 gallons ($516.7 - 160.7 = 356$).

77. c. To find the greatest margin of profit, start by looking at the graph and seeing which years have a greater distance between the two lines. Remember that profit is found by taking the money brought in (charge) and subtracting the cost. It looks like 2002 or 2003 could work so now we can calculate the actual profit. 2002—the charge was $145 and the cost was $95 for a profit of $50 ($145 - 95 = 50$). 2003—the charge was $150 and the cost was $105 for a profit of $45. Therefore 2002 has the largest margin of profit.

78. d. There is no information about how many cars were actually serviced. Even if the price is higher in one year, maybe not as many cars needed a tune-up so we can not assume anything for this question.

79. d. In 2001 the profit per car was $35 ($120 - 85 = 35$), so if $9,590 was profit, divide the profit by the profit per car ($\frac{9,590}{35} = 274$).

80. c. When mixing products together to end up with a new product, you must take into account the contribution of the parts. Cashews cost $5.25/lb, but we do not know how many lbs (x) are being mixed, so cashews can be represented by $5.25x$. The pre-made mixture costs $3.25/lb and there are 5 lbs of it, so the mixture can be represented by $3.25(5)$. These first two parts are being added together to make a new mixture worth $4.50/lb, and there will be a total of $5 + x$ lbs in the end (pre-made mixture + cashews). This information can be put into an equation:

$$5.25(x) + 3.25(5) = 4.50(x + 5)$$
$$5.25x + 16.25 = 4.5x + 22.5$$
$$\underline{-16.25 = -16.25}$$
$$5.25x = 4.5x + 6.25$$
$$\underline{-4.5x = -4.5x}$$
$$.75x = 6.25$$
$$x = 8.33$$

81. **a.** Volume is calculated by multiplying the length, height, and width together. The new length is 10″ (16 − 3 − 3 = 10). The height is determined by the squares that are removed—3″. The volume is 300 in³ (3 × 10 × 10 = 300).

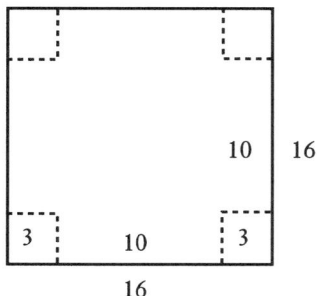

82. **a.** If you look at the diagram, the difference in surface area has to do with the corners. The only surfaces that the new piece is missing are from the corners. Each corner has 9 in² of area (3 × 3 = 9) and there are four corners, so the difference is simply 36 in² (9 × 4 = 36).

83. **a.** Dana covers d doctors and Faith has $\frac{2}{3}$ more than Dana, which is $\frac{5}{3}d$ ($d + \frac{2}{3}d \rightarrow \frac{3}{3}d + \frac{2}{3}d \rightarrow \frac{5}{3}d$). The equation to solve takes the total doctors and adds together Dana and Faith's coverage:

$$d + \frac{5}{3}d = 192$$
$$\frac{3}{3}d + \frac{5}{3}d = 192$$
$$\frac{3}{8} \times \frac{8}{3}d = 192 \times \frac{3}{8}$$
$$d = 72$$

84. **b.** This shape can be set up as two triangles that have the same area, and a rectangle. The rectangular piece has an area of 54 ft² (6 × 9 = 54). One of the triangles has an area of 13.5 ft² ($\frac{1}{2}$ × 9 × 3 = 13.5) so the two triangles have a combined area of 27 ft² (13.5 × 2 = 27). Add up all of the parts to find the total (27 + 54 = 81 ft²).

85. **b.** To find the perimeter we need to find out how long the four unknown sides are. This can be found by using the Pythagorean theorem on one of the triangles because one leg is 3 and the other is 4.5 ($\frac{9}{2}$ = 4.5). The hypotenuse is 5.4 from the following equation:

$$3^2 + 4.5^2 = c^2$$
$$9 + 20.25 = c^2$$
$$\sqrt{29.25} = \sqrt{c^2}$$
$$5.4 = c$$

There are four of these lengths in the hexagon and two that are 6′ long for a total of 33.6′ (4(5.4) + 2(6) = 33.6′).

86. **d.** This problem requires the use of the Pythagorean theorem ($a^2 + b^2 = c^2$). There is a right triangle formed by the support, the ground, and the wall. The hypotenuse is known, as is one of the legs. Plug in the values and solve for b.

$$(9.4)^2 + b^2 = (16.5)^2$$
$$88.36 + b^2 = 272.25$$
$$-88.36 = -88.36$$
$$b^2 = 183.89$$
$$\sqrt{b^2} = \sqrt{183.89}$$
$$b = 13.56$$

87. **b.** Profit is calculated by finding how much money came in, minus the cost. In this problem it will be helpful to have c = number of chairs Mary bought in the first place, and $c - 20$ = number of chairs sold. From the sales, Mary brought in $64(c - 20)$ and the original purchase of the chairs cost $38(c)$. So the equation to solve is:

$$64(c - 20) - 38(c) = 358$$
$$64c - 1,280 - 38c = 358$$
$$26c - 1,280 = 358$$
$$+ 1,280 = + 1,280$$
$$\frac{26c}{26} = \frac{1.638}{26}$$
$$c = 63$$

Therefore, she must have purchased 63 chairs in the beginning.

88. **d.** Matt needs a profit of $3,870 and he has purchasing costs of $2,900 ($100 \times 29 = 2,900$), for a total of $6,770 ($3,870 + 2,900 = 6,770$) that he needs to collect from sales. He is guaranteed to sell at least 80 faucets, so the money should be distributed among the 80 faucets by dividing ($\frac{6,770}{80} = 84.625$, round up to the nearest penny). Even though the exact value is not on the list, choose the price that is close and a little above to guarantee the profit.

89. **b.** To find the number of cars sold in 2002, find out how many cars/month all of them sold in 2002 ($33 + 42 + 36 = 111$). There are 12 months in a year so they sold approximately 1,332 cars ($111 \times 12 = 1,332$).

90. **b.** There were sales of 95 cars per month in 2001 ($28 + 32 + 35 = 95$), 111 cars per month in 2002 ($33 + 42 + 36 = 111$), and 114 cars per month in 2003 ($40 + 38 + 36 = 114$). The largest jump in sales took place in 2002, with an average of 16 more cars sold each month than in 2001.

91. **d.** Becky's increase goes up by 3 from 2000–2001, 5 from 2001–2002, and 7 from 2002–2003. It is reasonable to see a pattern and hope that she will increase by 9 in the coming year for a monthly total of 49 cars.

92. **c.** The computer is marked up 15% from $1,850 ($1,850 \times 1.15 = \$2,127.50$) and then discounted 25% ($2,127.50 \times .75 = \$1,595.63$). The last piece is to add the tax ($1,595.63 \times .065 = 103.72$) to the total price ($1,595.63 + 103.72 = \$1,699.35$).

93. **c.** The total number of pieces can be represented by $6x + 3x + x = 1,400$ because of the ratio that is held. $6x$ represents the number of 6-foot pieces, $3x$ represents the number of 8-foot pieces, and x represents the number of 12-foot pieces. Combine like terms to get $10x = 1,400$, so $x = 140$. Therefore, there are $3(140) = 420$ eight-foot pieces.

94. c. We are finding the surface area of a box with dimensions of 30 in (2.5 × 12 = 30) × 8 in × 3 in. There are six sides to a box — Front/Back, Side/Side, Top/Bottom. The front surface area is 90 (30 × 3 = 90), the side area is 24 (8 × 3 = 24), and the top area is 240 (30 × 8 = 240). The total of the front, side, and top is 354 (90 + 24 + 240 = 354) and all that is left is to double it to find the rest of the box (354 × 2 = 708).

95. b. Find out how many rooms Lacey can clean in one hour and add it together with how many rooms Maria can clean in one hour. The total time to clean is 1 hour = 60 minutes. Lacey can clean 4.29 rooms ($\frac{60}{14}$ = 4.29) and Maria can clean 3.16 rooms ($\frac{60}{19}$ = 3.16). When you add their efforts together they can clean 7.45 rooms, but because it asks for complete rooms, you must round down to 7 rooms.

96. b. The container is $\frac{2}{5}$ full, which means $\frac{3}{5}$ of the container is empty. The empty part of the container holds 217.5 m³. Using an equation, it is simple to find out what the total volume the container holds. Translate the following: $\frac{3}{5}$ of the container is 217.5 →

$$\frac{3}{5} \times V = 217.5$$
$$\frac{5}{3} \times \frac{3}{5} \times V = 217.5 \times \frac{5}{3}$$
$$V = 362.5$$

The total volume is 362.5 m³, so in order to find out how much is already in the container, subtract the empty volume from the total volume (362.5 − 217.5 = 145).

97. c. The volume of a cylinder is calculated from the following equation: $V = \pi r^2 h$. We know the volume, we will use 3.14 to approximate π, and with a quick calculation we will know the radius. The diameter is given, remember that the radius is half of the diameter, so the radius is 4.25 ($\frac{8.5}{2}$ = 4.25). Plug in the values to the equation and solve for the height.

$$362.5 = (3.14)(4.25)^2 h$$
$$362.5 = (3.14)(18.063)h$$
$$\frac{362.5}{57.72} = \frac{57.72h}{57.72}$$
$$6.28 = h$$

98. d. For a customer who only talks 140–155 minutes, Option 1 would only cost $7 per month. Option 2 would cost the same customer between $9.80 and $10.85 (.07 × 140 = 9.80 and .07 × 155 = 10.85). For someone who never talks more than 155 minutes per month, Option 1 is a better deal.

99. b. This is a proportion problem disguised with confusing information. The 2 : 5 ratio must be held in a proportion similar to this, where x represents the real height of the house:

$$\frac{2}{5} = \frac{10.2}{x}$$
$$2x = 51$$
$$x = 25.5$$

100. a. The pebble walkway is acting like a border around the pool. There is now a larger rectangle around the pool with an additional 5 m on each side (2.5 + 2.5 = 5). If you find the area of the larger rectangle and subtract away the area inside, you are left with the area of the border. It is almost like cutting out the middle of the big rectangle. The area of the large rectangle is 3,400 m² (*l* × *w* = *A*, 85 × 40 = 3,400). The area of the pool is only 2,800 m² (80 × 35 = 2,800). Subtract the two areas to find that the walkway is 600 m² (3,400 − 2,800 = 600).

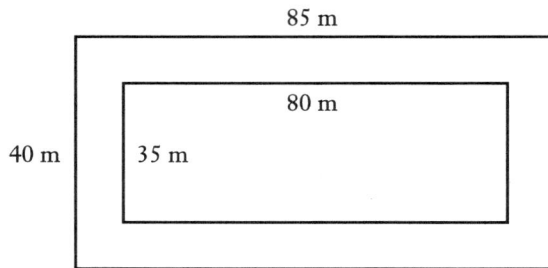

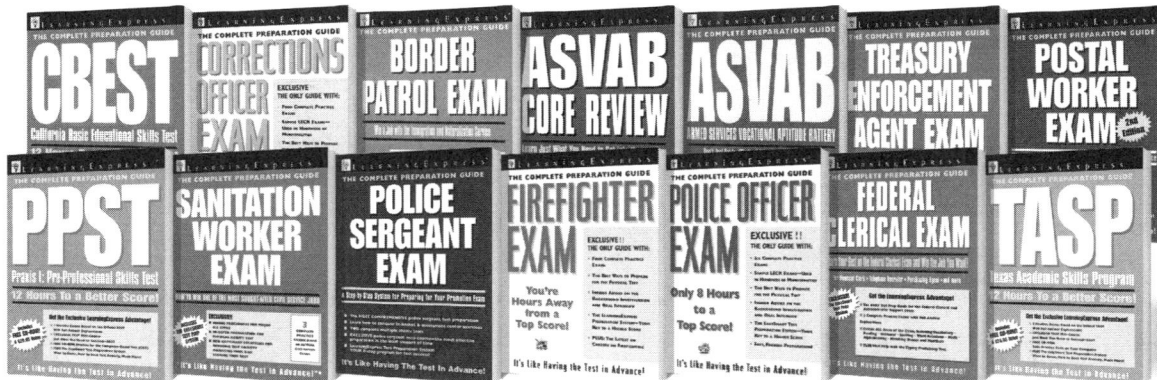

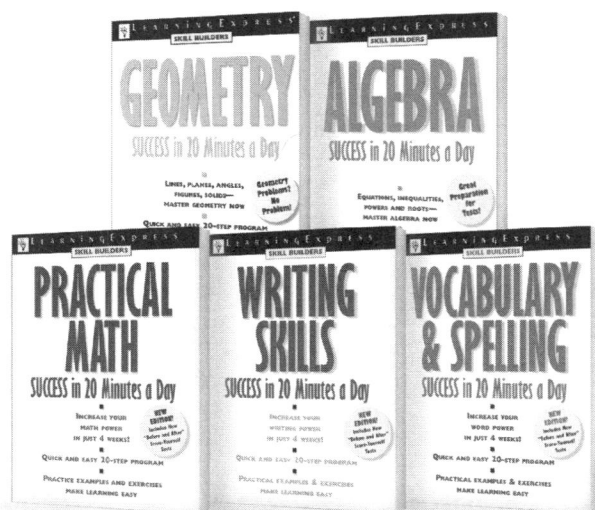